Southern Space Studies

The Southern Space Studies series presents analyses of space trends, market evolutions, policies, strategies and regulations, as well as the related social, economic and political challenges of space-related activities in the Global South with a particular focus on developing countries in Africa and Latin America. Obtaining inside information from emerging space-faring countries in these regions is pivotal to establish and strengthen efficient and beneficial cooperation mechanisms in the space arena and to gain a deeper understanding of their rapidly evolving space activities. To this end, the series provides transdisciplinary information for a fruitful development of space activities in relevant countries and cooperation with established space-faring nations. It is, therefore, a reference compilation for space activities in these areas.

More information about this series at http://www.springer.com/series/16025

Annette Froehlich

Editor

Embedding Space in African Society

The United Nations Sustainable Development Goals 2030 Supported by Space Applications

 Springer

Editor
Annette Froehlich (iD)
Space Lab, Department of Electrical
Engineering
University of Cape Town
Rondebosch, South Africa

ISSN 2523-3718   ISSN 2523-3726 (electronic)
Southern Space Studies
ISBN 978-3-030-06039-8   ISBN 978-3-030-06040-4   (eBook)
https://doi.org/10.1007/978-3-030-06040-4

Library of Congress Control Number: 2018964937

This Springer imprint is published by the registered company Springer Nature Switzerland AG
The registered company address is: Gewerbestrasse 11, 6330 Cham, Switzerland

Preface

Space continues to hold a fascination, a power, in our lives. For most people, that fascination is about the unknown, the mysterious secrets of the cosmic realm, which seems to whirl about far above us, removed from the Earthly concerns of daily life. Little do many people realise, or fully appreciate however, how close space truly is to their daily lives, both in value and importance, and in distance. In daily life, it is hard to find any activity which does not in some way rely on a fundamental level of space, its assets, or its data. Modern communications utterly depend on the satellites which fill the sky; banking, finance, mobile phones and their applications, including the Global Positioning System, weather forecasts, disaster relief, agriculture, climate observation, and countless other aspects of our lives exist as they do only because of space. In distance, low Earth orbit is no further away from each of us than London from Paris, or Washington DC from New York.

Thus, the power of space lies in the often subtle ways it makes modern life and its conveniences possible. But more than that, its power lies in its unfettered reach —space, being so close to every point on Earth, is also near to those places where traditional means cannot go without great difficulty. From the perspective of low Earth orbit, at first glance the planet appears undivided, undisturbed by human activity. Yet it was precisely this perspective which enabled us, towards the late twentieth century, to begin understanding the impact we were having on the planet, and how fragile it truly is. The need for sustainable development became clear, as was our utter dependence on the planet for our survival. Thus, while our activities in the last two centuries or so have transformed our lives entirely and granted us conveniences and affordances unimaginable to our ancestors, it was not without a cost to the environment.

Moreover, despite these the great achievements of economic and social development which have characterised recent history, and which have lifted many millions of people out of poverty, there is still much to be done if nobody is to be left behind, or denied the dignity of having their most basic needs met, including clean water, food, health care, education, peace, decent work, a sustainable and healthy habitat, and reduced inequalities. However, knowing now the importance of approaching these needs in a sustainable way, the member states of the United Nations came together in 2015 to commit themselves to improving the lot of the world's most vulnerable people in the form of the 17 Sustainable Development

Goals. Every effort must be made to fulfil these goals globally, and especially in Africa, where the some Millennium Development Goals—the precursor to the SDGS—went ruefully unmet despite great progress that was nevertheless achieved.

Because of the power of space and its applications in our daily lives, they play an undeniably important role in supporting the achievement of the SDGs. This is also recognised by the United Nations Office for Outer Space Affairs, which recently established a Working Group to focus on "Space 2030". While the importance of space in supporting the sustainable development objectives is equally applicable to Africa as it is for the rest of the world, woefully little has been published about the embedding of space into African societies in support of the SDGs. This publication is thus a timely contribution to the discussion around achieving the SDGs, and its particular focus on Africa fills an important gap in the discourse. Given the urgency of not leaving any of the SGDs unmet on the continent, it is critical to analyse the role played by space in this regard and to identify areas where space applications can further be brought to bear. By doing so, we stand the best chance of affording everyone on Earth the basic dignity they deserve, and of raising the profile, and appreciation, of space.

Rondebosch, South Africa   Dr. Annette Froehlich

Contents

Maximising the Use of Space Applications in Implementing the Sustainable Development Goals in Africa

Val Munsami

Abstract

The socio-economic challenges on the African continent are enormous and to tackle these challenges in a sustainable manner requires a systematic approach that cuts across a number of dimensions. There needs to be a stronger emphasis on transformative social and economic policies; whilst securing effective governance and democratic rule are preconditions for attaining regional stability, building confidence and reaching Africa's aspirations. Global Agendas such as Agenda 2030 and Agenda 2063, if appropriately adopted, become the guiding lens for implementing the socio-economic imperatives of Africa. This expose briefly looks at the underlying context, both historical and current, and connects the African aspirations, as enshrined in the African Agenda 2063, with the Sustainable Development Goals, as reflected in the global Agenda 2030. Core to the optimal achievement of these Agendas is the need for context relevant data. A case is made for spatial data, packaged as space applications services and products, to inform evidence-based policy making and to track the implementation progress of these Agendas. This expose further motivates for an African space programme based on the recently approved African Space Policy and African Space Strategy but cautions that the governance framework for the African Union proposed African Space Agency must be appropriately configured to service the needs of the African user community.

Keywords

Agenda 2063 · Agenda 2030 · African Space Policy · African Space Strategy · African space programme · Socio-economic challenges · Sustainable development goals

V. Munsami (✉)
South African National Space Agency, Silverton PO Box 484, 0127, Gauteng, South Africa
e-mail: vmunsami@sansa.org.za

© Springer Nature Switzerland AG 2019
A. Froehlich (ed.), *Embedding Space in African Society*, Southern Space Studies,
https://doi.org/10.1007/978-3-030-06040-4_1

1

1   Introduction

Any effective government administration has two fundamental national priorities that it must optimise, namely (i) the quality of life of its citizens, and (ii) wealth creation; through the adoption and implementation of relevant policy instruments. These complementary priorities essentially form opposite sides of the same coin that must be carefully balanced, and which collectively are referred to as "socio-economic development". On the one side, government plays a pivotal role in the macroeconomy of a country by mediating through a set of fiscal, monetary and growth policies to balance the competing interest of households (society), firms (private sector), government (public sector) and the international sector. On the other side, government must balance these macroeconomic interests with the optimal provision of public goods, i.e. societal benefits such as education and health care, which is defined as nonrival in consumption and nonexcludable.[1] A critical input factor in these deliberations is the optimum development and use of our natural resources. This introduces the notion of sustainable development, which is defined as the management of our natural resources to meet the immediate needs of the present population and the requirements of future generations.[2]

A key question is what sets developed countries apart from developing countries in terms of socio-economic development? At the core is the dynamic and complex inter-relationship between labour, natural resources and capital as input factors to national economies, but moreover how these economies are characterised and positioned in the global marketplace, and how well it responds to the social needs of the country. Peter Drucker[3] advocated that knowledge has become *the* resource, rather than *a* resource that makes our society 'post-capitalist', thus introducing the notion of a "knowledge economy", which fundamentally changes the structure of society, both in terms of economic and social dynamics. Policy makers must also account for the concomitant effect these dual dynamics have on the sustainability of our natural environment. The economies of developing countries are still largely labour and resource intensive and the notion of a knowledge economy and its associated social and environmental impacts have yet to be fully embraced. For this transition to happen, knowledge must be embedded and used for shaping constant change and organised for the systematic abandonment of historic norms and practices that hitherto has hindered progress, whether for social upliftment, economic transformation, political reform, or sustainable development. Innovation and entrepreneurship thus become the engine of sustained economic growth and social development within a knowledge economy.[4]

---

[1]Case K. E. and Fair R. C. (2001), *Principles of Economics*, 6th Edition, Prentice Hall: New Jersey [381, 331].
[2]Rao U. R. (1999), *Space Technology for Sustainable Development*, Tata McGraw-Hill Publishing Company Limited: New Delhi.
[3]Drucker, P. F. (1993), *Post-Capitalist Society*, Oxford: Butterworth Heinemann.
[4]Hill, C. W. L. (2014), *International Business: Competing in the Global Marketplace*, 10th Global Edition, McGraw-Hill: Glasgow.

The biggest barriers to developing knowledge economies on the African continent has been the lack of policy directives that commit resources to building such economic systems.[5] This problem is exacerbated when poor countries are held back by a short-term focus on poverty alleviation that constrains resources for future programs.[6] Even more problematic is the significant dependence on foreign aid by many African countries, which only serve to detract the long-term economic growth and developmental agendas of nation states. In addition, foreign direct investment (FDI) inflow is so critical for some African countries that it has become a significant component of fiscal planning and has consequently become a major driver for economic growth in these countries.[7] However, empirical evidence demonstrates that if countries are in transition to FDI openness, anticipated welfare gains lead to temporary declines in domestic investment and employment.[8] Hence, inappropriate fiscal and monetary policies can have adverse effects in shaping the economy and certainly on the quality of life of a nation. On the flip side, inappropriate social policies, for example policies relating to education that are not purposed to develop a knowledge workforce, could equally stifle the transition towards a knowledge economy.

Hence, effective policy making to constructively impact socio-economic development of a nation or region is evidently an exercise of counterbalancing social and economic policies in harmony with the environment. The singular factor that sits at this nexus is empirical data that could be used to differentiate between causal and acausal dependencies amongst a suite of dynamic variables that is then used to inform evidence-based policy making. In this chapter, we look at the socio-economic imperatives at the core of Africa's sustainable development by focusing on the developmental issues; expound on the broader socio-economic development policy landscape; consider what is required to unlock the constraints to effective policy making for socio-economic development; reflect on the role space applications can play in advancing sustainable socio-economic development; and then propose how this could be embedded at a regional level. We undertake this expose specifically through a policy lens with the appreciation that the cohesive glue for regional integration is understanding the spatial-temporal nuances and the inter-connectivity between national and regional socio-economic agendas.

---

[5]Munsami, V. and Nicolaides, A. (2017), Investigation of a Governance Framework for an African Space Programme, *Space Policy*, available online.

[6]Chataway, J., Smith, J. and Wield, D. (2006). Science and technology partnerships and poverty alleviation in Africa, *International Journal of Technology Management and Sustainable Development*, 5(2), 103–123.

[7]Adeleke, A.I. (2014). FDI-Growth nexus in Africa: Does governance matter? *Journal of Economic Development*, 39(1), 111–135.

[8]McGrattan, E. (2011). Transition to FDI Openness: Reconciling Theory and Evidence, Federal Reserve Bank of Minneapolic, *Research Department Staff Report*, 454, 1–44.

2   Socio-Economic Imperatives for Africa's Development

The aggregate socio-economic statistics for Africa are rather daunting. Africa currently comprises 55 countries, which collectively account for only 2.4% of the global Gross Domestic Product (GDP) even though it carries 30% of the world's mineral resources and is the fourth largest producer of oil. However, the African economic outlook has demonstrated resilience against the global financial crisis and real output growth is estimated to have increased to 3.6% in 2017, up from 2.2% in 2016, and projected to accelerate to 4.1% in 2018 and 2019.[9] The African Development Bank has recommended that African countries need to (i) lift their economies to a new growth equilibrium that is fuelled by innovation and productivity rather than by natural resources, (ii) ensure that fiscal policy does not undercut public investment that is important for securing the delivery of public goods, such as health and education, (iii) industrialise to end poverty and to generate employment for the roughly 12 million new entrants who seek to join its labour force annually, (iv) increase the stock of productive infrastructure such as power, water and transport services, which will cost between $130 billion and $170 billion a year, (v) take advantage of projected labour shortages elsewhere in the world that would require technological advances, and (vi) rapidly build skills in sciences, information and communications technology, engineering, manufacturing, and mathematics in order to participate in the fourth industrial revolution (i.e. the digital revolution that is core to building the services industry). Hence, the key to unlocking Africa's hidden potential is to drive economic growth and social emancipation through transforming local economies to knowledge economies.

The future economic prognosis, contrasted against historical performance, show signs of promise for Africa if the appropriate fiscal and monetary policy instruments are adopted. However, the social context and developmental aspirations must be factored into the shaping of these instruments to ensure a balanced approach to socio-economic development. For example, with respect to food security, how does Africa respond to its rising population, which is projected to double to 2.3 billion people by 2050; overcome chronic undernourishment of some 240 million Africans; and deal with the fact that over 90% of Africa's soil is unsuitable for agriculture. Likewise, how does Africa respond to global climate change with appropriate adaptation and mitigation measures when it is the world's hottest continent with one third being desert; when Africa is deemed to be the worst affected by climate change given the already widespread disease burden and rampant poverty; and Africa is the lowest contributor to greenhouse gas emissions whereas the global agenda is set by mainly developed nations, some of which have rejected climate change as a hoax. This problem set of food insecurity and the adverse impacts of climate change are intimately linked and have a fair degree of

---

[9]African Development Bank (2018), *African Economic Outlook 2018.*

co-dependency. Policy makers, therefore, cannot afford to develop policies to deal with one problem that works counter to resolving the other. This necessitates that socio-economic developmental issues be dealt with in a systematic fashion instead of adopting a piecemeal approach.

The aforesaid demonstrates the need to counterbalance economic and social policies, and that this is a complex multivariate multidisciplinary exercise. What is needed is a comprehensive quantifiable framework that defines the aspirations of the African continent, which collectively define the drivers for effective policy making. Such a framework currently exists in the form of Agenda 2063, where seven fundamental aspirations have been laid down, namely[10]:

1. A prosperous Africa based on inclusive growth and sustainable development,
2. An integrated continent, politically united and based on the ideals of Pan-Africanism and the vision of Africa's Renaissance,
3. An Africa of good governance, democracy, respect for human rights, justice and the rule of law,
4. A peaceful and secure Africa,
5. An Africa with a strong cultural identity, common heritage, shared values and ethics,
6. An Africa whose development is people-driven, relying on the potential of African people, especially its women and youth, and caring for children, and
7. Africa as a strong, united, resilient and influential global player and partner.

Each of these aspirations have been expanded to incorporate a set of indicators and targets that defines the Africa we aspire towards by 2063.

The primary institution that has been tasked with coordinating the implementation of Agenda 2063 is the African Union (AU). At the highest level, the common African agenda is espoused in the vision of the AU, namely:

"An integrated, prosperous and peaceful Africa, driven by its own citizens and representing a dynamic force in the global arena".

The vision of 'an integrated, prosperous and peaceful Africa' is certainly not attainable without the appropriate endogenous scientific, engineering and technical capabilities being developed and harnessed for this common goal. In addition, any effort toward the realisation of this vision will be underpinned by the strength and solidarity of Africa's political, economic, social, and environmental commitments, which will ultimately determine its collective success and the relevance thereof in the global arena.

---

[10]African Union (2016), *Agenda 2063: The Africa We Want*, Popular Version - May 2016 Edition: African Union, Addis Ababa, Ethiopia.

3 Historical Pitfalls to Africa's Socio-Economic Development

To fight the war against poverty and underdevelopment in Africa, significant resources, including capital, technology and human skills, need to be mobilised and utilised into productive areas needed for development. Underpinning the mobilization of these critical resources is the need for government effectiveness, political stability, regulatory quality, rule of law and the absence of corruption, which are determinants for accelerating the ease of doing business in African countries.[11] Good governance institutions guarantee economic freedom by providing growth-enhancing incentives, which helps define the investment environment and creates the favourable conditions needed for economic growth. A holistic approach to development is required, which stresses the linkage between security, good governance and economic development and that such development can only be achieved in a secure and democratic environment, conducive to long-term invest-ments.[12] This approach relates governance to the delivery of high quality political goods, where these political goods relate to safety and security, rule of law, par-ticipation and human rights, sustainable economic opportunity and human development.[13]

In this regard, African countries have had a mixed governance record with countries being bogged down with political instability, government ineffectiveness, the lack of rule of law, and serious problems with corruption, all of which are signs of poor governance.[14] This has led to state fragility that is characterised by a society that is fractured, the economy being mismanaged, and social service delivery is severely deficient such that the social contract between the state and its citizens has failed.[15] When such fragility sets in then political leadership is weakened to a point that systematic terror, rape, property destruction, population displacement and the forced conscription of the youth tend to become the norm. This indefensible context becomes fertile ground for criminal syndicates and illegal activities. This state of anarchy has two attendant effects, namely, (i) neighbouring countries suffer eco-nomically due to regional instability, and (ii) international security is threatened. It has been demonstrated that neighbouring states suffer on average a 2% decline in

---

[11]Alemu, A.M. (2013). The nexus between governance infrastructure and the ease of doing business in Africa, *International Journal of Global Business*, 6(2), 34–56.

[12]Bagoyoko, N. and Gibert, M.V. (2009). The linkage between security, governance and development: the European Union in Africa, *Journal of Development Studies*, 45(5), 789–814..

[13]Rotberg, R.I. (2009). Governance and leadership in Africa: Measures, methods and results, *Journal of International Affairs*, 62(2), 113–126.

[14]Fayissa, B. and Nsiah, C. (2013). The impact of governance on economic growth in Africa, *Journal of Developing Areas*, 47(1), 91–108.

[15]Rugumamu, S.M. (2011). Capacity Development in Fragile Environments: Insights from parliaments in Africa, *World Journal of Entrepreneurship, Management and Sustainable Development*, 7(2/3/4), 113–175.

economic growth rates as a consequence of regional instability.[16] On the other hand, following the September 2001 attack on the World Trade Centre in New York, fragile states were seen to incubate international terrorism and became a mammoth focus of Western national security agencies, which has played out in countries such as Liberia and Somalia[17].

Given the aforementioned challenges at the level of individual States, the regional challenges are by no means trivial. At a broad socio-political level there are two intervening factors that negate regional unity. Firstly, of the 55 African countries, only two countries, namely Liberia and Ethiopia, were not colonised. This scramble for territorial control resulted in Africa being divided and ruled within boundaries that ignored the broader ethnic, linguistic, cultural and religious makeup of the native people on the continent. The resulting humiliation and inhuman treatment started a movement for the independence of African States, which was given further impetus when colonisation attempts of the last standing bastions failed.[17] This resulted in the establishment of the Organisation of African Unity in 1963, which has subsequently morphed into the African Union of today. Although colonial rule has ended, neo-colonialism is still active, where countries are still politically and economically reliant on their past colonial rulers. The second factor is the linguistic divide, primarily as an outcome of the first factor, where the official languages adopted by African countries as a medium of communication, are Arabic, English, French and Portuguese. These linguistic heritages also have profound political complications; with Arabic communities politically linked through common cultural identities across North Africa and the Middle East; and Anglophone, Francophone and Lusophone countries, respectively, adopting the cultural and political undertones of their western complements. In addition, the social fabric of Africa is complex and where national pride and tensions prevail the cultural differences play out to diminish regional unity.

A further dimension that has hindered regional unity is the issue of effective negotiations underpinned by the modes of regional governance in Africa. Governance at a regional level amounts to systems of rule in which goals are pursued through the exercise of control and for which three modes of regional governance have been identified, namely[18]:

1. Neoliberal regional governance—founded on the notion that regional economic integration should be market driven and outward looking and should remove obstacles to free trade within the region;

---

[16]Rugumamu, S.M. (2011). Capacity Development in Fragile Environments: Insights from parliaments in Africa, *World Journal of Entrepreneurship, Management and Sustainable Development*, 7(2/3/4), 113–175.

[17]Alemayehu, M. (2000). *Industrializing Africa: Development options and Challenges for the 21st Century*, Africa World Press, Inc.: Asmara, Eritrea.

[18]Soderbaum, F. (2004). Modes of Regional Governance in Africa: Neoliberalism, Sovereignty Boosting, and Shadow Networks, *Global Governance: A Review of Multilateralism and International Organizations*, 10:4, 419–436.

2. Sovereignty boosting governance—state steered regional governance can be portrayed as a means to promote national interest, particularly postcolonial states which tend to be obsessed with absolute sovereignty; and

3. Shadow governance—which relates to informal market activities with the purpose of promoting aid networks and private self-interests, and which is prevalent in fragile States.

These modes of regional governance often work against each other and pose a significant barrier to regional agenda setting and progress. Neoliberal and sovereignty boosting governance will be the norm at the negotiation table, as these represent definitive drivers for change and the promotion of self-interests. In the case of neoliberal politics, regional blocs negotiate to optimise the economic benefits to a specific region, whereas in the case of sovereignty boosting the primary interest is that of the nation State alone. Shadow governance, however, remains a particular challenge due to the fact that a number of African states are not democratically governed, and, in many instances, an authoritarian leader's sole interest is surviving power.[19] However, under regionalisation and globalisation pressures, these fragile States are dissolving with the development of a global political economy and economic liberalisation.[20] Because of these competing agendas, regional disunity is evident and often leads to an absence of collective action. These problems are especially evident when reflecting on the financial commitments of African Member States to the AU. The financial balance sheet of the AU is largely dominated by external donor funds and as such Member States do not feel obliged to undertake or commit to constructive action at the continental level. Unless there is collective ownership and resources committed to championing its own destiny, Africa will otherwise fall foul to the crazes of external agendas.

4   Relevance of the Sustainable Development Goals to Africa

Due to the challenges summarised above, the need for socio-economic development on the African continent becomes an even higher priority. The significant lag experienced by African countries compared to their developed counterparts, to a large extent relates to governance and effective leadership in adopting sound socio-economic policies to spur a knowledge-based economy that in equal part addresses the social challenges of a country or the region. When a suite of geo collocated countries faces similar socio-economic challenges, there are two important dimensions that need to be considered and balanced, namely (i) giving

---

[19]Welz, M. (2013). The African Union beyond Africa: Explaining the limited impact of Africa's continental organisation on global governance, *Global Governance*, 19, 425–441.

[20]Sandu, C. and Haines, R. (2014). Theory of governance and social enterprise, *The USV Annals of economics and public administration*, 14(2), 204–222.

precedence to national priorities, and (ii) fostering regional neoliberal collaboration. Whether acting severally or collectively, by adopting similar aspirations and policies, countries are able to promote regional integration, share common experiences and best practices, and tackle regional or global challenges that transcend national boundaries. This is where global agendas like the Agenda 2030 find relevance and mutual expression through the identification of a set of shared challenges, and associated metrics for evaluating progress, to unlock the key determinants underpinning socio-economic development.

In 2015, senior representatives from 193 countries convened at the United Nations headquarters in New Work to adopt a new developmental agenda called the 2030 Agenda for Sustainable Development or commonly referred to as the Sustainable Development Goals (SDGs). This was a mammoth achievement under the headship of the United Nations Development Programme, where the global community adopted 17 sustainable development goals that are tailored to end poverty, hunger, inequality, tackle climate change, and build resilient societies whilst taking care of our natural environment. These goals have quickly emerged as a global blueprint that all tiers of society, whether local or global, can associate and resonate with. It is instructive to note that both Agenda 2063, adopted in January 2015, and Agenda 2030, adopted in December 2015, were independently intended as a post-2015 follow on from the Millennium Development Goals (MDGs), adopted in 2000 and extending up until 2015. The MDGs, ratified by 191 countries, contained eight goals, namely[21]:

1. To eradicate extreme poverty and hunger,
2. To achieve universal primary education,
3. To promote gender equality and empower women,
4. To reduce child mortality,
5. To improve maternal health,
6. To combat HIV/AIDS, malaria, and other diseases,
7. To ensure environmental sustainability, and
8. To develop a global partnership for development.

Over its 15-year lifespan, the MDGs provided a useful framework for driving targeted interventions relating to socio-economic development, in an environmentally sustainable manner. From an African perspective, commendable progress was made on five of the eight MDGs, namely Goals 2, 3, 4, 6, and 7; whilst progress on Goals 1, 5 and 8 were slow.[22] Whilst significant progress has been made globally towards achieving the goals and targets of the MDGs, the drive for socio-economic development was by no means complete. This acknowledgement spurred a global effort that started in 2012 during the United Nations Conference on Sustainable

---

[21]End Poverty: Millennium Development Goals and Beyond 2015. Accessed from http://www.un.org/millenniumgoals, 2018-07-31.

[22]African Union, United Nations Economic Commission for Africa, African Development Bank and United Nations Development Programme (2016), *MDGs to Agenda 2063/SDGs: Transition Report*, ECA Documents Publishing Unit: Addis Ababa, Ethiopia.

Development in Rio de Janeiro, Brazil, to develop a follow on to the MDGs, which ultimately resulted in the adoption of Agenda 2030 in 2015. In preparation for the development of the SDGs, a Common African Position (CAP) on the post-2015 development agenda was developed.[23] In adopting the CAP, the Heads of State of the AU emphasised "That the post-2015 development agenda provides a unique opportunity for Africa to reach consensus on common challenges, priorities and aspirations, and to actively participate in the global debate on how to provide a fresh impetus to the MDGs and to examine and devise strategies to address key emerging development issues on the continent in the coming years. The post-2015 development agenda should also reaffirm the Rio Principles, especially the principle of common but differentiated responsibilities, the right to development and equity, and mutual accountability and responsibility, as well as ensure policy space for nationally tailored policies and programmes on the continent…".

It is important to note that Agenda 2030 was heavily influenced by the CAP, as Africa was the only region to have submitted a position in writing that became a working document within the United Nations Open Working Group and which position was also used for inter-governmental negotiations.[24] Thus, Africa foresaw a need to bring to the table a common African position to influence the global discourse that shaped and framed Agenda 2030, which demonstrates the intrinsic relevance of the SDGs to Africa's socio-economic development. At this juncture it is essential to draw a parallel between the global Agenda 2030 and the African Agenda 2063. Whereas Agenda 2030 comprises of 17 SDGs and 169 targets, Agenda 2063 comprises of 20 Goals and 174 targets. The respective Goals of Agenda 2030 and Agenda 2063 is shown in Table 1. There are three principal differences between the Goals of Agenda 2030 and Agenda 2063. Firstly, the lifespan of Agenda 2030 is 15 years, whereas that of Agenda 2063 is 50 years segmented into five 10-year implementation periods. Thus, whereas Agenda 2030 is a medium-term framework, Agenda 2063 is a long-term framework to be implemented in distinct phases. Secondly, there are four Goals of Agenda 2063 that have no correlation with Agenda 2030 and is specific to the pressing needs of Africa, namely, (i) a united Africa (federal or confederate), (ii) continental financial and monetary institutions established and functional, (iii) African cultural renaissance is pre-eminent, and (iv) a fully functional and operational African peace and security architecture. Thirdly, Agenda 2030 is a framework to be adopted at a national level and where there is an expectation for countries to submit National Reports. On the other hand, Agenda 2063 is a regional framework and for which the reporting on progress is meant to provide relative inter-national comparisons to enable targeted interventions and support.

It is important to observe that all 17 SDGs find expression in one or more Goals of Agenda 2063, which further demonstrates the relevance of Agenda 2030 to Africa's socio-economic development. However, from a governance perspective

---

[23]African Union (2014), *Common African Position*.
[24]African Union (2018), *Agenda 2063-SDGs*, Accessed from https://au.int/en/ea/statistics/a2063sdgs, 2018-07-31.

Table 1 The respective Goals of Agenda 2063 and Agenda 2030

Goals of Agenda 2063	Goals of Agenda 2030
1. A high standard of living, quality of life and well-being for all citizens	1. End poverty in all its forms everywhere
2. Well educated citizens and skills revolution underpinned by science, technology and innovation	2. End hunger achieve food security and improved nutrition and promote sustainable agriculture
3. Healthy and well-nourished citizens	3. Ensure healthy lives and promote well-being for all at all ages
4. Transformed economies	4. Ensure inclusive and quality education for all and promote lifelong learning opportunities for all
5. Modern agriculture for increased productivity and production	5. Achieve gender equality and empower women and girls
6. Blue/ocean economy for accelerated economic growth	6. Ensure availability and sustainable management of water and sanitation for all
7. Environmentally sustainable and climate resilient economies and communities	7. Ensure access to affordable, reliable, sustainable and modern energy for all
8. A United Africa (Federal or Confederate)	8. Promote sustained, inclusive and sustainable economic growth, employment and decent work for all
9. Continental financial and monetary institutions established and functional	9. Build resilient infrastructure, promote sustainable industrialisation and foster innovation
10. World class infrastructure criss-crosses Africa	10. Reduce inequality within and among countries
11. Democratic values, practices, universal principles of human rights, justice and the rule of law entrenched	11. Make cities inclusive, safe, resilient and sustainable
12. Capable institutions and transformative leadership in place	12. Ensure sustainable consumption and production patterns
13. Peace, security and stability is preserved	13. Take urgent action to combat climate change and its impacts
14. A stable and peaceful Africa	14. Conserve and sustainably use the oceans, seas and marine resources
15. A fully functional and operational African peace and security architecture	15. Sustainably manage forest, combat desertification, halt and reverse land degradation, halt biodiversity loss
16. African cultural renaissance is pre-eminent	16. Promote peaceful and inclusive societies for sustainable development and provide access to justice for all
17. Full gender equality in all spheres of life	17. Strengthen the means of implementation and revitalise the global partnership for sustainable development
18. Engaged and empowered youth and children	
19. Africa as a major partner in global affairs and peaceful co-existence	
20. Africa takes full responsibility for financing her development	

this poses a challenge given the capacity and resource constraints in Africa to pursue multiple instruments. The solution to this problem is finding the intersecting points between these instruments and to develop an integrated planning and reporting tool to harmonise the adoption of Agenda 2063 and Agenda 2030 into national planning frameworks and to reduce the transaction costs of reporting on both.[25] This is precisely what the United National Economic Commission for Africa has done through the development of an integrated toolkit, which requires enactment of the following recommendations for its effective implementation[25]:

1. Leverage synergies among multiple development initiatives by mapping the relationships between multiple development initiatives.
2. Institutionalize platforms for advocacy and awareness to promote buy-in of all stakeholders.
3. Integrate the different agendas into national development plans.
4. Invest in and strengthen capacities for the use of appropriate planning tools.
5. Strengthen capacities for policy simulation to identify inter-sectoral synergies and trade-offs and shape policy prioritisation and sequencing.
6. Strengthen national statistics offices and plans.
7. Partnership and coordination with development partners.

Such an integrated toolkit is vitally important to Africa's socio-economic development, as it not only reduces the transaction cost of reporting on both Agenda 2030 and Agenda 2063 but provides an effective bridge between the regional priorities enshrined in Agenda 2063, and how this finds contextual relevance to the global Agenda 2030 at a national level. Moreover, the adoption of the SDGs at a national level provides a reciprocal bridge where progress on Agenda 2030 likewise contributes to the regional aspirations of Agenda 2063.

5   Relevance of Space Applications to Africa's Socio-Economic Development

Depending on their mission, satellites have different orbits. Weather and communication satellites are placed in Geostationary Orbits (GEO) at an altitude of 36,000 km above the equator, from which they have a constant gaze on the same hemisphere of the earth. These satellites complete one orbit around the earth every 24 hours. Other satellites are placed in Low Earth Orbits (LEO), which complete on average one orbit around the earth every 100 min. Because the earth rotates across the plane of the orbit, such a satellite eventually covers the whole earth. Such orbits are used for remote sensing, navigation and positioning, and space exploration applications. It should be noted that these satellites are a means to a broader end, where the end is the development of appropriate products and services that (i) supports our way of life, (ii) monitors the state of our environment, and (iii) provides vital information for evidence-based decision making.

---

[25]United Nations Economic Commission for Africa (2017), *Integrating Agenda 2063 and 2030 Agenda for Sustainable Development into national development plans*, ECA Documents Publishing Unit: Addis Ababa, Ethiopia.

Earth observation/remote sensing satellites use modern instruments to gather information about the nature and condition of earth's land, sea, and atmosphere. Located in various orbits, these satellites use sensors that can "see" a broad area and report very fine details about the weather, the terrain, and the environment. The sensors receive electromagnetic emissions in various spectral bands that show objects, which are visible, such as clouds, hills, lakes, and many other features. These instruments can detect an objects temperature and composition, the wind's direction and speed, and environmental conditions, such as erosion, fires, and pollution.[26] Remote sensing satellites have thus become a formidable weapon against the destruction of the environment, because they can systematically monitor large areas to assess the spread of pollution and other damages.

Satellite communications is a key technology that could effectively enable developing countries to participate in the build-up of the global information infrastructure. Wireless systems are the most cost-effective way to develop or upgrade telecommunications networks in areas where user density is lower than 200 subscribers per square kilometre. Such wireless systems can be installed 5–10 times faster and at a 50% lower cost than landline networks.[27] Technologies for education and training, in particular distant education and multimedia, may be instrumental in meeting the needs of developing countries that have to train and integrate a large number of workers in widely dispersed and under-equipped areas. Many countries have to cope with large-scale disease outbreaks and telemedicine may help to meet these challenges by improving the organisation and management of health care through tele-consultation and the support of remote medical assistance. At a glance one can tell which parts of the country are clear or cloudy. When satellite maps are put in motion we easily see the direction of clouds and storms. Countless lives are saved every year by this simple ability to track the paths of hurricanes and other deadly storms. By providing farmers with valuable climatic data and agricultural planners with precise information, this technology has improved food production and crop management.

Satellite navigation uses satellites as reference points to calculate positions on earth accurate to within a few meters. With advanced techniques and augmentation systems, satellite navigation can provide measurements down to within a meter. Navigation and positioning receivers have been miniaturized and are becoming economical thus making the technology accessible to everyone. For example, Global Navigation Satellite Services (GNSS) receivers are currently built into cars, boats, planes, construction equipment and even laptops.[28] Navigation and positioning are the main elements of the international air traffic management system providing worldwide navigation coverage to support all phases of flight. In general, mariners use the GNSS for either navigation or positioning on the high seas and this

---

[26]UNISPACE III (1999). *Management of earth resources*. Office for Outer Space Affairs, Document A/CONF.184/BP/3.

[27]UNISPACE III (1999). *Space communications and applications*. Office for Outer Space Affairs, Document A/CONF.184/BP/5.

[28]UNISPACE III (1999). *Satellite navigation and location systems*. Office for Outer Space Affairs, Document A/CONF.184/BP/4.

has recently been applied to the surveillance of illegal shipping activities, such as fisheries and monitoring oil spills. Many automotive navigation and positioning applications fit within the description of intelligent transportation systems (ITS) and we are now seeing the emergence of self-driven vehicles that rely on ITS.

Hence, since the dawn of the space age, about five decades ago, we have come to rely heavily on space technologies and its associated products and services. Daily weather forecasts, instantaneous worldwide communications, and a constant ability to record high-resolution images are all examples of space technologies that we have come to depend upon. Even basic commodities, such as food and energy resources, are facilitated by space-based technology. In fact, the high standard of living in developed countries is largely attributed to the adoption and application of space-based technologies.[29] This convenient lifestyle is supported by the instant access to information and space-based applications, such as GNSS and global television coverage. Of late, there has been an increasing appreciation and use of the value proposition of space applications towards addressing socio-economic challenges and responding to opportunities, as space-derived services are deemed crucial for socio-economic development. For example, approximately 70–80% of the Essential Climate Variables used for monitoring climate change are derived from satellite data.

Africa is also slowly awakening to the realisation and appreciation of the impact that space science and technology can make in addressing its socio-economic challenges. However, the benefits of these services have accrued to Africa indirectly, as a consumer of services and products provided by multi-national companies and inter-governmental agencies. While some of these products and services have helped to serve the social and economic needs of the continent, Africa cannot boast of possessing the technical know-how to participate effectively in space-related activities. This external dependency limits Africa in developing an endogenous space sector, which is a high-end technology sector that has a myriad of direct and indirect spinoffs. In this respect, Africa presents a significant growth potential, especially given its disparate socio and economic lag compared to other developed regions of the world. What is therefore needed is a formal programme that is characterised by self-determination and self-sufficiency.

From a policy relevance perspective, the AU is mandated to coordinate Africa's response to Agenda 2063, and also Agenda 2030, through the adoption of the CAP. The AU comprise of eight Commissions that are delegated to oversee critical policy issues that emphasise the main tenets that are fundamental to achieving Africa's aspirations, and these Commissions are:

1. Political Affairs,
2. Economic Affairs,
3. Social Affairs,
4. Peace and Security,

---

[29]MacPhail, D. (2009). Increasing the use of earth observation in developing countries, *Space Policy*, 25(1), 6–8.

5. Trade and Industry,
6. Infrastructure and Energy,
7. Human Resources, Science and Technology, and
8. Rural Economy and Agriculture.

It is insightful to determine the reliance of these Commissions on space technology and applications in achieving their set mandates. Table 2 maps out the eight Commissions together with the broad objectives for each of these. For each of these objectives, an indication is given as to whether space science and technology is needed in the implementation of the said objectives. As can be seen all of the 40 objectives require the use of or would be greatly assisted by space science and technology for its achievement. It therefore makes intuitive sense that investments

Table 2 Relevance of space applications to the work of the AU

Department of Political Affairs		*Department of Economic Affairs*	
Democracy and electoral assistance	✓	Economic policies and research	✓
Democracy, governance, human rights and elections	✓	Economic integration and regional cooperation	✓
Humanitarian affairs, refugees and displaced persons	✓	Private sector development, investment and resource mobilisation	✓
		Statistics division	✓
Department of Social Affairs		*Department of Peace and Security*	
Health, nutrition and population	✓	Common African defence and security policy	✓
HIV/AIDS, malaria, tuberculosis	✓	Operationalisation of continental peace and security architecture	✓
Labour, employment and migration	✓	Support efforts to prevent, manage and resolve conflicts	✓
Social welfare, vulnerable groups and drug control	✓	Coordination, harmonisation and promotion of peace and security	✓
Culture and sport	✓	Post conflict resolution and development	✓
		Structural prevention of conflicts	✓
Department of Trade and Industry		*Department of Infrastructure and Energy*	
Intra-inter-regional trade	✓	Harmonise ICT policy and regulation	✓
Industrial development and diversification of Africa's economy	✓	Development of the ICT sector	✓
Increased and freer trade to foster economic growth	✓	Equitable, accessible, affordable and innovative ICT services	✓
Department of Human Resources, Science and Technology		*Department of Rural Economy and Agriculture*	
Education	✓	Agriculture and Food Security	✓
Human resources and youth	✓	Environmental and natural resources	✓
Science and technology	✓	Rural economy	✓

in space applications and associated technologies should form the mainstay in achieving the aspirations of Agenda 2063, and by implication Agenda 2030, and by association the vision of the AU. Hence, this provides compelling evidence for the relevance of space applications to Africa's socio-economic development.

Gauging from the reliance of the AU Commissions on space science and technology in meeting their priorities and the need for addressing Africa's socio-economic challenges, it is therefore apparent that space applications, and the associated platforms, have a critical role to play on the continent. Most importantly, space applications provide the effective tools, in many domain areas, for evidence-based policy making. Evidence based policy making is extremely important for Africa's development as it means the difference between advancing indigenous solutions to local problems versus adopting solutions from elsewhere in the world that in many instances are largely inappropriate, especially given the differing social contexts.

6 Foreseen Challenges to the Implementation of the SDGs in Africa

Counterbalancing social and economic policies that jointly are in sync with the long-term sustainability of the environment requires evidence-based policy making. The estimation of the potential impact related to specific policy choices is an exercise in forecasting future trajectories based on historical performance. Also, drawing from the MDG experience, an important lesson learnt is that the initial conditions influence the pace of progress a country or region can make in achieving its developmental agendas. In this respect, how do we draw on recent technological advances that enables nowcasting, which is a trigger for Big Data and the 4th Industrial Revolution. The MDG experience also "exposed the data challenges facing national statistical systems and underscored the importance of strengthening statistical and analytical capacities. The data requirements for tracking progress on the SDGs will be greater than those for the MDGs, reflecting the SDGs' broader scope and the emphasis on disaggregation of data. Confronting this challenge will require strengthened human and financial capacities, together with new approaches and methodologies for harnessing the wealth of information which can be obtained from big data".[30] This experience draws attention to three intertwined factors that are needed for the effective implementation of the SDGs in Africa, namely:

1. An appropriate governance architecture, at national and regional levels, to drive the implementation of the SDGs and the reporting thereon.
2. Adequate resources, both human and financial, that are committed for the implementation of national or regional priorities that are aligned with the SDGs.

---

[30]United Nations Economic Commission for Africa (2015), *MDG Report 2015: Assessing Progress in Africa toward the Millennium Development Goals*, ECA Documents Publishing Unit: Addis Ababa, Ethiopia.

3. Spatial datasets that are needed both for evidence-based policy making and for reporting on progress of the SDGs.

Whilst the governance and resourcing factors are critical for the implementation of the SDGs in Africa, fit for purpose data becomes the glue that elucidates, and binds evidence-based policy making with the ability to track and respond to socio-economic challenges; whilst enabling an effective reporting mechanism. Thus, improved data fuels better governance and informs the appropriate allocation of resources, while data gathering focuses attention on the situation being analysed and reported on.[31]

The responsibility for reporting on the SDGs at a national level sits with the statistical agencies. However, implementation of the SDGs resides with other public and private sector institutions. This fact and the need to shift from national to a regional level for the seamless reporting on Agenda 2030 and Agenda 2063, brings to the fore the exceeding importance of open data policies. Currently, there is a strong motivation for full, free and open access to public sector digital data so that the benefits associated from such public investments is maximised.[32] The potential benefits from simply publishing public data as open data include the following[33]:

- The publicising of data will automatically yield benefits;
- All public information will be publicised without restriction;
- Every constituent can make use of the open data; and
- Open data will result in open and integrated governance.

However, caution must be raised with respect to these potential benefits, namely, that availability of open data in itself does not imply that the data will be used. Central to this concern is the capacity and need to utilise this data. This problem is certainly exacerbated on the African continent where the appreciation of the value of data is lacking and the capacity to use such data is sub-optimal.

While open data is relatively new for society at large, some parts of the space sector have been accustomed to the open data practice. For example, astronomers have considered open access as beneficial for advancing the knowledge and understanding of the discipline.[34] The open access debate is especially important for satellite earth observation. As the scope of the data products and services has increased from satellite-based platforms, the associated information has expanded accordingly to include earth science products into the mainstream environmental,

---

[31]Alexander, G. and Endres, J. (2014). *The trouble with statistics in Africa*, Accessed from https://africacheck.org/2014/10/28/comment-the-trouble-with-statistics-in-africa/, 2018-07-31.

[32]Harris, R. and Baumann, I. (2015). Open data policies and satellite earth observation, *Space Policy*, 32, 44–53.

[33]Janssen, M., Charalabidis, Y. and Zuiderwijk, A. (2012). Benefits, adoption barriers and myths of open data and open government, *Information Systems Management*, 29, 258–268.

[34]Harris, R. and Baumann, I. (2015). Open data policies and satellite earth observation, *Space Policy*, 32, 44–53.

meteorological and other user communities.[35] A prime example of such use is the inclusion of GNSS satellite signals for global navigation services in the transportation sector. This highlights the need for robust space-based applications, given the need to have access to comprehensive data to document the dynamics of a changing environment.[36] For these reasons, it is critical that an open data policy is adopted on the African continent that will ensure the equitable sharing of satellite-based data given the inherent public good associated with such datasets.[37] Also, drawing from South Africa's experience, as the second country, after Mexico, to table a Country Report on progress made with respect to implementation of the SDGs, the Statistician General found difficulty in knowing exactly what data was needed and where the source of such data was located. These challenges certainly need to be dealt with within the governance architecture for reporting on Agenda 2030 and Agenda 2063.

Given the important role space applications play in addressing socio-economic challenges, it is useful to reflect on the current state of the African space sector. The growth of space activities on the African continent has largely been limited to a few countries that have initiated national space programmes (and in some cases space agencies). These countries include Algeria (Agence Spatiale Algerienne—ASAL), Egypt (National Authority for Remote Sensing and Space Sciences—NARSS), Ghana (Ghana Space Science and Technology Institute—GSSTI), Kenya, Nigeria (National Space Research and Development Agency—NASRDA) and South Africa (South African National Space Agency—SANSA). A few more countries are investigating the option of setting up national space programmes and these include Angola, Botswana, Ethiopia, Mauritius, Namibia, Tanzania and Zambia. However, the collective state of space activities among all these Member States is still at an infancy level, with South Africa being the most advanced yet only being formally initiated in 2011 amidst changing policy drivers.[38] Thus, the general state of the local adoption and implementation of space applications on the African continent varies considerably from one country to another. This reality reflects the lack of a general political will and appreciation to embrace space science and technology as a driver for socio-economic development.

A comparative assessment of the relative global expenditure in space related activities, shown in Fig. 1, highlights the lack of commitment to advancing space science and technology by countries in Africa. South Africa (ranked 23rd) is the only African country that ranks in the top 30 countries in terms of a committed

[35]Gilruth, P.T., Kalluria, S., Robinson, J.W., Townshend, J., Lindsay, F. and Davis, P. (2006). Measuring performance: moving NASA earth science products into the mainstream user community, 22, 165–75.

[36]Brown, M.E., Escobar, V.M., Aschbacher, J., Milagro-Perex, M.P., Doorn, B., Macauley, M.K. and Friedl, L. (2013), Policy for robust space-based earth science, technology and applications, *Space Policy, 29*, 76–82.

[37]Harris, R. and Miller, L. (2011). Earth observation and the public good, *Space Policy, 27*, 194–201.

[38]Munsami, V. (2014). South Africa's national space policy: The dawn of a new space era, *Space Policy, 30*, 115–120.

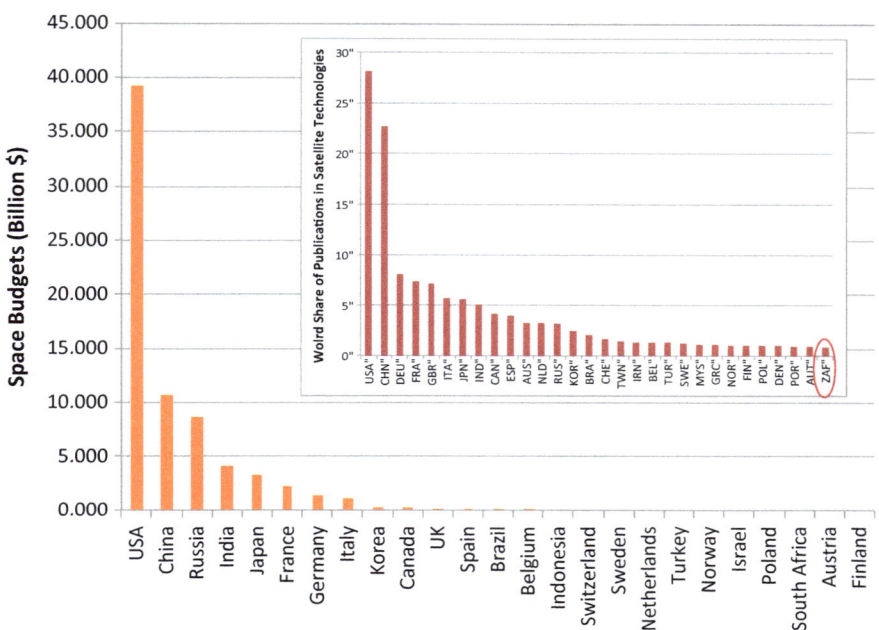

Fig. 1 Global expenditure and publication rates in satellite engineering (Munsami, V. (2014). South Africa's national space policy: The dawn of a new space era, Space Policy, 30, 115–120.)

budget for space activities. Even this is skewed as the majority of the funding (more than 80% of the budget) is spent in radio astronomy, particularly for the global Square Kilometre Array project. South Africa is also the only country that ranks in the top 30 performers in terms of publication output for satellite engineering (ranked 30th), as shown in the insert of Fig. 1. In addition, even though Africa represents 20% of the earth's land surface, more than the USA, India, China and Europe put together, yet these countries/regions spent more than $50 billion on space activities in 2013 compared to less than $100 million spent by the African continent in the same period.[39] These comparisons highlight significant short-comings in terms of under-investments and suboptimal activities in the space sector, which limit Africa's potential in a fast-growing sector that is vital to addressing its socio-economic challenges. However, at this juncture, it is important to ask whether Africa is being reactive to global trends or if there is a genuine need for an African space programme. Many critical commentators from the global North have questioned whether Africa should be investing in space science and technology when it has so many social ills to tend to. The implication therefore, is that Africa will do better to address its social challenges than invest in such a capital-intensive domain. This unfortunately is a lax opinion and is uninformed from a Science, Technology and Innovation (STI) policy position, which advocates that STI is a driver for social

---

[39]OECD (2014). *Space Economy at a Glance 2014.* OECD Publication.

reform and an engine for economic growth. Therefore, to effectively address Agenda 2030 and Agenda 2063, Africa must develop an effective STI system, inclusive of an African space programme, that responds to these Agendas.

7 Embedding Space Applications Within Africa

Given the above reality, work was begun in 2013 in developing the requisite instruments, namely an African Space Policy and an African Space Strategy, needed to define and shape an indigenous African space programme. These instruments were approved by the AU Heads of States in January 2016 and provide a framework that encapsulates a common vision and direction for Africa in capturing the benefits of space applications in addressing its manifold socio-economic challenges. The contextualisation of an appropriate governance architecture is embedded within these instruments, which architecture must be established to ensure effective implementation of an African space programme that meets the needs of its citizens. The African Space Policy reflects the core principles, associated with a set of objectives, which are needed to inform the activities to be carried out within a formal African space programme.[40] It provides the guiding framework by which the relevant actors on the continent are encouraged to follow and which adherence will promote a systemic effect thus leading to a well-coordinated and integrated African space programme. Such a systemic approach will ensure convergence towards the main tenets enshrined within the African space strategy that will ultimately assist in responding to the major socio-economic challenges on the continent. The Policy, in essence, provides the governance protocol for a formal African space programme and thus forms the bedrock for such institutionalisation.

The strategic approach for the African Space Strategy is to adopt a user pull philosophy where relevant user requirements, that are reliant on space applications, are responded to. In the process of attempting to address user requirements, recognition must be given to the current technology options available in the continental space landscape. These options provide a basis from which future technology needs should stem from. Hence, the user requirements provide the broad parameters within which the appropriate technology platforms, both current and new, will emerge and evolve. In this regard, a user requirements assessment was conducted, as part of the development of the African Space Strategy, to identify the primary use of space applications. The high-level outcome of this exercise together with a mapping against the four space thematic areas, namely, (i) earth observation, (ii) navigation and positioning, (iii) satellite communications, and (iv) space science

---

[40]African Union (2017), *African Space Policy*, Accessed from https://au.int/sites/default/files/newsevents/workingdocuments/33178-wd-african_space_policy_-_st20444_e_original.pdf, 2018-07-31.

and astronomy, is shown in Table 3.[41] In particular, the spatial resolution (i.e. the minimum object size that can be resolved) and the temporal resolution (i.e. the revisit time) for earth observation are also shown. It is clear that the thematic foci of earth observation, navigation and positioning, and satellite communications are required for the majority of user requirements. It should be noted, that the identified user requirements, in many instances, is also an extended expression of the needs of the AU Commissions. These user requirements are also embedded at national levels and effectively form the drivers for national space programmes. Table 4 provides a high-level mapping of these user requirements against the SDGs, which is by no means meant to be a comprehensive cross-correlation but only a succinct reflection of the relevance of these user requirements to the SDGs.

The success of the formal African space programme will, therefore, be primarily measured by the extent to which the technology platforms, and the services and products meet these user requirements and by association the SDGs. It should be noted that the inherent value of a satellite platform is not equivalent to the capital related cost, but instead on how effective it is in meeting the original mission requirements. For example, a $50 million-dollar satellite will be worth nil if it does not deliver the input data it was designed to produce for the associated services and products. Supporting programmes are critical for the realisation of the services and products that respond to the various user requirements. The supporting programmes essentially give effect to the policy objectives and principles of the African Space Policy and the strategic directives enshrined in the African Space Strategy. In this regard, the supporting programmes are the critical readiness factors that are needed to make the formal space programme a successful and sustainable long-term endeavour. These supporting programmes include:

1. **Human capital and space awareness**—the appropriate expertise and skills necessary for a space programme is an area that must receive priority attention, as without them all existing and envisaged programmes and infrastructure will be of little value. In order for the formal space programme to be meaningful to the broader public, there will certainly be a need for creating public awareness relating to the benefits that space technology and its manifold application products and services can deliver.
2. **Infrastructure**—appropriate infrastructure is the cornerstone of an effective space programme, which enables technology transfer and human capacity development initiatives. Such infrastructure must adequately cover the full value-chain of activities needed for the space mission concept. Optimal use must be made of infrastructure that currently exists on the continent and where it is lacking then upgrades and new builds must be pursued.
3. **International partnerships**—strategic partnerships with foreign partners are necessary for tangible and intangible technology transfer and maintaining a viable and sustainable space programme. Such partnerships are vital for creating

---

[41]African Union (2017), *African Space Strategy*, Accessed from https://au.int/sites/default/files/newsevents/workingdocuments/33178-wd-african_space_strategy_-_st20445_e_original.pdf, on 2018-07-31.

Table 3 African user needs mapped against the space thematic areas

User needs	Earth observation											Navigation and positioning	Satellite	Space science and astronomy
	Spatial resolution								Temporal resolution					
	<50 cm	50 cm–1 m	1–2.5 m	2.5–5 m	5–10 m	10–20 m	20–30 m	>30 m	Daily	Seasonal	Annual			
Disasters	✓	✓	✓	✓	✓	✓	✓	✓	✓			✓	✓	✓
Health					✓	✓				✓		✓	✓	✓
Energy				✓	✓	✓					✓	✓	✓	✓
Climate				✓	✓	✓			✓			✓		✓
Water		✓	✓	✓	✓	✓	✓	✓		✓		✓	✓	
Weather		✓	✓	✓	✓	✓	✓	✓	✓			✓	✓	✓
Ecosystems				✓	✓	✓	✓	✓		✓		✓		
Agriculture				✓	✓	✓	✓	✓	✓			✓	✓	
Biodiversity				✓	✓	✓	✓	✓			✓	✓		
Peace, safety and security	✓	✓	✓		✓			✓	✓			✓	✓	✓
Human migration and settlements		✓	✓	✓							✓	✓	✓	
Education and human resources				✓	✓	✓	✓	✓			✓	✓	✓	✓
Communications				✓	✓	✓	✓	✓				✓	✓	✓
Trade and industry			✓	✓	✓	✓	✓	✓		✓		✓	✓	
Transport		✓	✓	✓	✓	✓	✓	✓			✓	✓	✓	
Infrastructure		✓	✓	✓	✓	✓		✓	✓			✓	✓	

Table 4 High-level mapping of user requirements against the SDGs

		African user requirements															
		Disasters	Health	Energy	Climate	Water	Weather	Ecosystems	Agriculture	Biodiversity	Peace, safety and security	Human migration and settlements	Education and human resources	Communications	Trade and industry	Transport	Infrastructure
Sustainable development goals	End poverty in all its forms everywhere		✓	✓	✓	✓							✓	✓			
	End hunger achieve food security and improved nutrition and promote sustainable agriculture				✓		✓	✓	✓	✓							
	Ensure healthy lives and promote well-being for all at all ages		✓														
	Ensure inclusive and quality education for all and promote lifelong learning opportunities for all												✓	✓			
	Achieve gender equality and empower women and girls												✓	✓			

(continued)

Table 4 (continued)

| | African user requirements | | | | | | | | | | | | | | | |
	Disasters	Health	Energy	Climate	Water	Weather	Ecosystems	Agriculture	Biodiversity	Peace, safety and security	Human migration and settlements	Education and human resources	Communications	Trade and industry	Transport	Infrastructure
Ensure availability and sustainable management of water and sanitation for all					✓											
Ensure access to affordable, reliable, sustainable and modern energy for all			✓													
Promote sustained, inclusive and sustainable economic growth, employment and decent work for all												✓	✓	✓		
Build resilient infrastructure, promote sustainable industrialisation and foster innovation													✓	✓		✓

(continued)

Table 4 (continued)

African user requirements

	Disasters	Health	Energy	Climate	Water	Weather	Ecosystems	Agriculture	Biodiversity	Peace, safety and security	Human migration and settlements	Education and human resources	Communications	Trade and industry	Transport	Infrastructure
Reduce inequality within and among countries												✓	✓			
Make cities inclusive, safe, resilient and sustainable		✓	✓		✓						✓				✓	
Ensure sustainable consumption and production patterns							✓		✓							
Take urgent action to combat climate change and its impacts	✓			✓		✓	✓									
Conserve and sustainably use the oceans, seas and marine resources				✓			✓									
Sustainably manage forest, combat desertification, halt and reverse land degradation, halt biodiversity loss	✓						✓		✓							

(continued)

Table 4 (continued)

	African user requirements															
	Disasters	Health	Energy	Climate	Water	Weather	Ecosystems	Agriculture	Biodiversity	Peace, safety and security	Human migration and settlements	Education and human resources	Communications	Trade and industry	Transport	Infrastructure
Promote peaceful and inclusive societies for sustainable development and provide access to justice for all	✓								✓							
Strengthen the means of implementation and revitalise the global partnership for sustainable development												✓	✓			

the necessary critical mass that will drive the formal space programme. In addition, when Africa is fully capacitated it will explore and pursue equal partnership initiatives that will assist in moving the technology boundaries for Africa's benefit.

4. **Industrial participation and development**—development of the regional space industry to participate in various space missions is a key requirement for the sustainability of a formal African space programme. Doing so not only benefits the space sector but helps promote and develop other sectors through spinoff benefits. Industry participation in the formal space programme is essential for developing and delivering the services and products both for the commercial market and for public good activities.

Lastly, an effective governance system needs to be instituted for the implementation of an African space programme. This process is well advanced with the AU formalising the Statutes for an African Space Agency and with the AU currently in the process of assessing a Host Country for the said Agency. However, the governance framework for this Agency must be carefully chosen, as any adverse decisions in this regard could have longer-term detrimental effects by limiting potential opportunities and reducing the impact of the Agency vis-à-vis its mandate. A few recommendation relating to the functions of an African Space Agency include[42]:

- The purpose of the Agency should be the promotion and coordination of space activities on the continent.
- The operating principle for the Agency should be a multilateral approach, where majority consensus defines the decision-making process.
- For the African space programme to be sustainable, two key challenges should be addressed, namely (i) ensure optimal political support, and (ii) reduce the dependency on external sources of funding.
- Regional relevance should be ensured by the establishment of regional satellite offices for the Agency and for the Agency to work directly with the Member States, where necessary.
- The development of space products and services, and space-based platforms should be coordinated centrally by the Agency.
- Ground-based facilities required for the African space programme should be co-managed between the Agency and the Member States that have been assigned to host such regional facilities.
- The development of the African space market and the African space industry should be the concern of all stakeholders straddling the Agency, the Regional Economic Communities and the Member States.

---

[42]Munsami, V. and Nicolaides, A. (2017), Investigation of a Governance Framework for an African Space Programme, *Journal of Space Policy*, available online.

- The Agency should engage with its users through the establishment of Communities of Practices for each of the identified user requirements and also engage directly with the users, where necessary.
- A common African position for multilateral engagements should be driven by the Agency.
- The Agency should coordinate a continent wide regulatory framework for space activities on the continent.
- The African Space Agency should work directly with the National Space Agencies when interfacing with the Member States and in the co-management of space activities for the continent.

8 Conclusions

Africa's socio-economic development is a gargantuan exercise in counterbalancing social and economic policies, but also to give due consideration to the political pitfalls to Africa's socio-economic development. This is especially apt given the poor governance record that has oft times resulted in political dictatorship and regional instability, and the neoliberal regional and sovereignty boosting governance approaches that works against regional unity. There is also a recognised need for African economies to transition to knowledge-based economies that is simultaneously purposed to tackle the intrinsic social challenges. This brings to the fore the importance of global Agendas such as Agenda 2063 and Agenda 2030, as frameworks that could be adopted to assess progress with respect to the socio-economic imperatives of Africa. However, to reduce the transaction costs associated with multiple reporting there is a need to adopt an integrated planning and reporting framework that connects Agenda 2030 with Agenda 2063.

This expose provides a compelling case of the need for space applications in implementing the SDGs and Agenda 2063 in Africa and the concomitant need for an African space programme. Core to the achievement of the SDGs and the goals of Agenda 2063 is access to context relevant data that both informs evidence-based policy making and tracking of the progress made with respect to these global Agendas. Spatial data should be applied in a systematic manner for evidence-based policy making to separate out causal and acausal dependencies amongst a suite of dynamic variables and then to successively correlate those with co-dependencies. In addition, aggregated spatial data should be made accessible to national statistical agencies in order to ensure efficient reporting on implementation of the aforementioned Agendas. However, to ensure that the African Space Agency is able to respond to the spatial data needs of the African user community, the governance framework for the Agency must be tailored to suite the African environment.

Author Biography

Valanathan Munsami holds a PhD in Physics and is currently the Chief Executive Officer of the South African National Space Agency (SANSA). He was involved in the development of South Africa's National Space Strategy and National Space Policy and oversaw the establishment of SANSA. He also Chaired the African Union Space Working Group that was tasked with the development of the African Space Policy and the African Space Strategy, which were approved by the African Union Heads of State in January 2016. He is currently a Member of the South African Council for Space Affairs (SACSA) and is a Vice President of the International Astronautic Federation for Developing Countries and Emerging Nations. He is also a Member of the International Academy of Astronautics.

Passive to Active Space and the Role of Space Assets in Sustainable Development

Samuel Anih

Abstract

The 2030 Agenda for Sustainable Development was designed to promote better living conditions for Earth and its inhabitants through the implementation of several goals that drive technological innovation and developments; these goals are also geared towards inclusive solutions for the populace and environmental protection for the long term preservation of Earth. The focus of the UN 2030 Agenda is to ensure that both developed and emerging countries benefit from the goals unlike the Millennium Development Goals which focused on developing countries. One of the prominent and viable contributors to these goals is the availability and the use of various space assets launched to space and the supporting infrastructure located on Earth. As a cutting edge technology, space assets are poised to promote human and sustainable development through their strategic and vantage point in space and the potential they have to contribute to human development while helping to preserve the planet. Africa as a continent requires the instrument and strategies contained in the SDGs to move forward. Few countries in Africa have space heritage from early Space Age while many others are currently active in the space arena reaping the benefit of space and its contribution to various aspects of national developmental programs with regards to several SDG goals. This chapter discusses the role of space assets for Africa in achieving the UN 2030 Agenda with a bias towards poverty reduction, prosperity and partnership in the SDGs.

Keywords

SDGs · Passive space · Active space · Partnership · Space assets

S. Anih (✉)
Spacelab, University of Cape Town, Cape Town, South Africa
e-mail: anihsammy@gmail.com

S. Anih
ARCSSTE-E, OAU Campus, Ile-Ife, Nigeria

© Springer Nature Switzerland AG 2019   31
A. Froehlich (ed.), *Embedding Space in African Society*, Southern Space Studies,
https://doi.org/10.1007/978-3-030-06040-4_2

1   Introduction

The Resolution 70/1—Transforming our world: the 2030 Agenda for Sustainable Development—with 17 goals, 169 targets and 231 indicators was adopted by the General Assembly on 25 September 2015 with a clear cut plan as contained in the declaration:

> We resolve, between now and 2030, to end poverty and hunger everywhere; to combat inequalities within and among countries; to build peaceful, just and inclusive societies; to protect human rights and promote gender equality and the empowerment of women and girls; and to ensure the lasting protection of the planet and its natural resources. We resolve also to create conditions for sustainable, inclusive and sustained economic growth, shared prosperity and decent work for all, taking into account different levels of national development and capacities.[1]

Consequently the objective set out to create prosperity for all with the shared responsibility by all to make the world a better place while preserving the planet. After the expiration of the Millennium Development Goals (MDGs) in 2015, the Sustainable Development Goals (SGDs) which have broader and more ambitious scope with the inclusion of developed nations. The 2030 Agenda for Sustainable Development was therefore conceived to be comprehensive with the potential to foster equitable global future for all if the goals are adequately implemented by all concerned. This makes the SDGs goals global unlike the MDGs which specifically focused on developing countries. Figure 1 shows the timelines and focus of various UN development goals.

It is expected that governments and concerned organizations integrate the goals into their programmes through policy and necessary platforms in order to allow for proper implementation through identification of required goal and setting of appropriate targets with results that are properly analyzed. This will ensure proper evaluation to ascertain if set targets are met. The evaluation of the outcomes of chosen targets as well as the attendant data comparison with appropriate indicators would allow for efficient governance that ensures accountability, transparency and policies beneficial to the humans and the planet alike.

Five key themes (5Ps) underpinning the 17 Sustainable Development Goals (SDGs) are described by the 2030 Agenda for Sustainable Development[2]: People, Planet, Prosperity, Peace and Partnership. The summary of the 5Ps are given below:

- *People: end poverty and hunger, in all their forms and dimensions, ensuring that all human beings can fulfil their potential in dignity and equality and in a healthy environment.*
- *Planet: protect the planet from degradation, including through sustainable consumption and production, sustainably managing its natural resources and*

---

[1]United Nations, 'Transforming Our World: The 2030 Agenda for Sustainable Development' (2015) A/RES/70/1 1.
[2]ibid.

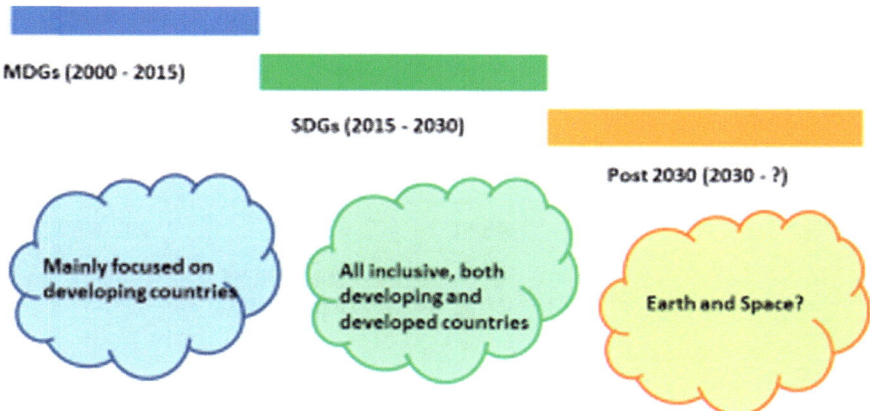

Fig. 1 Timelines and focus of past, present and possible future development goals. Samuel Anih, 'Earth and Extra-Terrestrial Sustainable Development: The Challenges of Post-2030 Earth and Space Regime' in Annette Froehlich (ed), *Post 2030-Agenda and the Role of Space* (Springer 2018)

> *taking urgent action on climate change, so that it can support the needs of the present and future generations.*

- ***Prosperity***: *ensure that all human beings can enjoy prosperous and fulfilling lives and that economic, social and technological progress occurs in harmony with nature.*
- ***Peace***: *foster peaceful, just and inclusive societies which are free from fear and violence. There can be no sustainable development without peace and no peace without sustainable development.*
- ***Partnership***: *mobilize the means required to implement this Agenda through a revitalized Global Partnership for Sustainable Development, based on a spirit of strengthened global solidarity, focused in particular on the needs of the poorest and most vulnerable, and with the participation of all countries, all stakeholders and all people.*[3]

One way of grouping the goals under the 5Ps would be to arrange them such that each of the five clusters encapsulates various goals contained in the 17 SDGs (Fig. 2).

- *People*: 1–6 (1-No Poverty, 2-Zero hunger, 3-Good health and well-being, 4-Quality education, 5-Gender equality, 6-Clean water and sanitation).
- *Prosperity*: 7–11 (7-Affordable and Clean Energy, 8-Decent work and economic growth, 9-Industry, innovation and infrastructure, 10-Reduce inequality within and among countries, 11-Sustainable cities and communities).

---

[3]ibid.

Fig. 2 The 5Ps of
sustainable development.
Wayne Visser, 'UN
Sustainable Development
Goals – Finalised Text and
Diagrams' (2015) <http://
www.waynevisser.com/tag/
sustainable-development>
accessed 9 April 2015

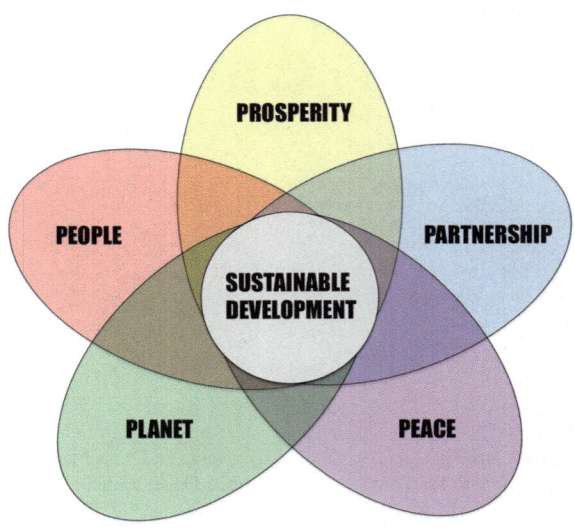

- *Planet*: 12–15 (12-Responsible consumption and production, 13-Climate action, 14-Life under water, 15-Life on land).
- *Peace*: 16 (16-Peace, justice and strong institutions).
- *Partnership*: (17-Partnerships for the goals).

2   Challenges of the African Continent

The African continent has a lot of challenges—from the prevalence of extreme poverty to lack of adequate infrastructural development. The level of under development in sub Saharan Africa is substantially higher when compared to other regions in other continents of the world. Despite the growth being recorded by countries in Africa there has not been commensurate improvement in the standard of living. Presently many countries in Africa are way behind other countries in the world and are not able to comparatively compete while still grappling with poor economic growth as well as low infrastructural development several years after independence. For example, many African countries gained their independence around the same time as Singapore but are currently playing the catch-up game in terms of economic development. Singapore as at 1965 (at independence) had a GDP per capita of $516.3 while countries in sub Saharan Africa had GDP per capita of $162.8, but currently Singapore has gone far ahead economically with a GDP of $57,714.3 in 2017 while that of Sub Saharan Africa is $1,553.8[4]. Data from World

---

[4]World Bank, 'World Bank Data - GDP per Capita (Current US$)' (2018) <https://data.worldbank.org/indicator/NY.GDP.PCAP.CD> accessed 18 July 2018.

Bank shows that despite the reduction in extreme poverty to about 41%, the continued increase in population means that 389 million people in the region lived on less than $1.90 a day in 2013, 113 million more than in 1990.[5] About one out of two Africans currently lives in extreme poverty even though it is expected to fall to between 16 and 30% by 2030.[6] This has implications in terms of economic development, job creation, social services, healthcare, housing, governance, well-being, prosperity, science and technology development and a lot more.

African countries have not fared better in the area of science and technology development. In a document prepared for the World Bank, *"Science and Technology Collaboration: Building Capacity in Developing Countries?"* RAND Corporation's Science and Technology Policy Institute ranked 150 countries on their potential to innovate and collaborate with more scientifically advanced nations using Index of Science and Technology. Mauritius was the highest ranking African nation at number 59 while out of 20 last countries on the list 14 were African.[7] An update of this index by Gayle Allard in 2015 shows an improvement even though the challenges with its science and technology output in Sub Saharan Africa still persist.[8]

This is of concern as there is a correlation between science and technology development and economic growth, also there is the need to improve on several contributing factors in this area. The benefits of science and technology depend on the following: human resource investment training and development, private sector demand for knowledge, provision of appropriate enabling environment by public policies for strong knowledge institutions as well as the standard of technologies systems used for information and communication that permit the flow and dissemination of knowledge and information.[9] Space science and technology is one of the ways countries can quickly innovate, develop and drive economic growth.

---

[5]World Bank, 'Atlas of Sustainable Development Goals 2017 : From World Development Indicators' (2017) <https://openknowledge.worldbank.org/handle/10986/26306> accessed 17 July 2018.

[6]Punam Chuhan-Pole and others, 'Economic Prospects for Sub-Saharan Africa Remain Strong, but Growth Is Vulnerable to a Sharp Decline in Commodity Prices', vol 8 (2013) <http://www.worldbank.org/content/dam/Worldbank/document/Africa/Report/Africas-Pulse-brochure_Vol8.pdf #page=21>.

[7]Gayle Allard and others, 'Science and Technology Collaboration: Builiding Capacity in Developing Countries.' [2001] Journal of African Studies and Development 137 <https://www.rand.org/content/dam/rand/pubs/monograph_reports/2005/MR1357.0.pdf>.

[8]Gayle Allard, 'Science and Technology Capacity in Africa: A New Index' (2015) 7 Journal of African Studies and Development 137 <http://www.academicjournals.org/article/article1434799458_Allard.pdf>.

[9]Robert Watson, Michael Crawford and Sara Farley, *Strategic Approaches to Science and Technology in Development. Policy Research Working Paper:No. 3026* (No 3026, World Bank).

3   Space on the African Continent

This section of the document discusses the activities of space in the African con-
tinent; this is explored under passive and active space, highlighting the continents
heritage with space and current trends in the sector.

3.1   Passive Space: Africa Space Heritage and Progress

With the advent of spaceflight on 4th October 1957, humankind have been able to
obtain better vantage point for exploration and exploitation of resources on Earth
and in space.

Participation in space programmes comes in two ways—passive and active
space. The passive space in this document describes space activities where the
participating or host country only contributes or lends a parcel of her territory for
building of facilities to be used by another country to further the latter's own space
programs.

Passive space activities that took place within the continent of Africa from the
1960s to 1970s ensured that Africa played significant role in contributing to the
success of other space programs at the beginning of the Space Age. Three
prominent ones will be discussed in this document.

3.1.1   Tracking Stations in Nigeria

During the early part of the space program there were efforts to launch spacecraft as
part of the International Geophysical Year. The Americans wanted to catch up with
the Soviet Union after the latter launched the first artificial satellite Sputnik-1 in
October 4 1957 followed by Sputnik-2 on 4 November 1957. The Explorer-1
satellite was successfully launched on January 31 1958 by the Americans but there
was the need to track the satellite, this led to setting up of series of tracking stations
around the globe. The US army had earlier deployed portable radio tracking stations
in Nigeria, Singapore and California. The stations were instrumental to the
telemetry received and assisted in plotting the spacecraft's orbit at mission control.[10] In fact it was the Nigerian tracking station that heralded the existence of Van
Allen Belts from Explorer's signals.[11] The success of the Explorer tracking station
in Nigeria paved way for other tracking stations to be established after an agreement
just two weeks after Nigeria's independence between Nigeria and the United States
leading to the construction of the NASA Tracking Station in the outskirts of Kano,
Northern Nigeria (Fig. 3).[12]

---

[10]NASA, 'NASA Facts—Deep Space Network' <http://www.jpl.nasa.gov/news/fact_sheets/DSN-0105.pdf>.
[11]National Academy of Sciences, 'Memorial Tributes' (2011) 14.
[12]Adigun Ade Abiodun, 'Nigeria's Destiny with Space' in Robert Ajayi Boroffice and Joseph
Akinyede (eds), *Nigeria's Quest in Space* (2013).

Fig. 3 A 1962 Photograph of the NASA tracking station in Kano, Nigeria used for tracking early American human spaceflights—from *"Read you loud and clear!: The story of NASA'S spaceflight tracking and data network"*. Sunny Tsiao, *Read You Loud and Clear! : The Story of NASA'S Spaceflight Tracking and Data Network* (The NASA H, NASA History Division, Office of External Relations 2008) <https://history.nasa.gov/STDN_082508_50810-20-2008_part1.pdf>

3.1.2 Tracking Station in Hartebeeshoek

Several tracking stations with bigger antennas were later established by NASA all over the world, among them were Ascension Island in the South Atlantic, Woomera and Tidbinbilla in Australia, Robledo and Cebreros in Spain, Goldstone in California, Cape Canaveral in Florida and Hartebeeshoek in South Africa. The 26 m antenna, Deep Space Station 51 (DSS-51) was installed near Hartebeesthoek outside Johannesburg in South Africa. This was used to first support Ranger 1's mission to the moon and to test systems for future missions. The tracking station at Hartebeeshoek also contributed to tracking space probes headed for the moon and others further away in the solar system such as the Pioneer Jupiter, its most distant target, which was designed to explore the interplanetary medium beyond the orbit of Mars, investigate the nature of asteroid belt and explore the Jovian environment.[13] Other missions tracked at Hartebeeshoek include Surveyor missions to the Moon, Lunar Orbiters, Apollo missions, Mariner missions to Venus and Mars.[14] The DSS-51 was operational without a break through the first decade and half of the Space Age.[15] But due to the changing flight requirements as well as political

---

[13]NASA, 'South African Antenna' (2013) <https://www.nasa.gov/directorates/heo/scan/services/networks/DSN50Gallery-16.html> accessed 15 July 2018.

[14]HartRAO, 'Missions Tracked at Deep Space Station 51 at Hartbeesthoek' (2012) <http://www.hartrao.ac.za/other/dss51/missions.html> accessed 20 August 2018.

[15]NASA (n 13).

Fig. 4 Deep Space Station 51 (DSS-51) was completed in 1961; it ceased operations in 1974. Sunny Tsiao, *Read You Loud and Clear! : The Story of NASA'S Spaceflight Tracking and Data Network* (The NASA H, NASA History Division, Office of External Relations 2008) <https://history.nasa.gov/STDN_082508_50810-20-2008_part1.pdf>

pressure on the Americans, the facility ceased operations in 1974 after which NASA equipment was taken away (Fig. 4).[16]

3.1.3   The San Marco Platform

Another country in Africa that contributed to the space activities through launching programmes is Kenya. The San Marco platform near the coast of Kenya was used for launching several satellites by foreign countries such as USA, UK and Italy. The location has the advantage of adding extra boost to launched rockets due to the faster rotation of the earth near the Equator. A notable satellite that was launched from this platform was the American 64-kg X-ray satellite named Uhuru - launched on 12 December 1970.[17] The facility provided access to space to various countries except Kenya on whose territorial waters it was located.

The above case studies of Nigeria, South Africa and Kenya provide a background/perspective into the nature of space participation from the early days of the Space Age up till the conclusion of the various space related programs by the hosted countries. Despite the vital roles played by the first two countries towards the success of various American space programs, they were immediately discontinued once the hosted facilities met their space program objectives. The use of the San Marco platform complex commenced in March 1964 and concluded in 1988.

---

[16]ibid.

[17]NASA-GSFC, 'The Uhuru Satellite' (2003) <https://heasarc.gsfc.nasa.gov/docs/uhuru/uhuru.html> accessed 7 July 2018.

Table 1 Satellites launched by African countries to date

	Country	Satellite launched	Year of first satellite launch
1	Algeria	Alsat-1, Alsat-1B, Alsat-2A, Alsat-2B, Alsat-1 N, AlcomSat-1	2002
2	Angola	AngoSat-1	2017
2	Egypt	Nilesat 101, NileSat 102, Nilesat 201, EgyptSat 1, EgyptSat 2[a]	1998
4	Ghana	GhanaSat-1	2018
5	Kenya	1KUNS-PF	2018
6	Morocco	Mohammed VI-A	2017
7	Nigeria	NigeriaSat-1, NigcomSat-1, NigeriaSat-2, NigeriaSat-X, Nigcomsat-1R Nigeria EduSat-1	2003
8	South Africa	Sunsat-1, Sumbandila,nSight-1, ZACUBE-1	1999

[a]NileSat 103 and NileSat 104 were leased from Hotbird 4 and Atlantic Bird 7 respectively

3.2 Active Space: Current Space Efforts in Africa

The last section highlighted three examples of passive space activities that took place on the African continent during the beginning of the Space Age. Several African countries continue to aspire to participate and benefit from space activities in order to maximize the benefit therein as global development catches on within the African continent. Few of African countries have their own national space programmes dating back to early 1990s while others like South Africa dates further back to 1947 with amateur rocketry that eventually metamorphosized into a full blown missile program by 1993[18] with establishment of a space agency in 2010. Few others have officially started their various space agencies and made their mark on the world stage by launching their own satellites to space. Table 1 lists African countries that have launched satellites to space, their satellites (some of them defunct) and the year their first satellites were launched.

Most African satellites are in low Earth orbit (Earth observation satellites) while a few others are in geo-stationary orbit (communication satellites).

3.3 Partnership in Space

Space age started as a spin-off from established military programmes designed by the then super powers to showcase military power and wield international influence. Even then, it has been recognized that space activities have a way of uniting nations, even those having political tensions back on Earth, a very good example is the relationship between United States and Russia where space activities between the two countries

---

[18]Keith Gottschalk, 'South Africa's Space Program' (2010) 8 Astropolitics 35.

still progressed even after the United States imposed economic sanctions on Russia after the latter's invasion of Crimea which resulted in strained relationship between both countries. Their partnership through the International Space Station, where astronauts from both countries live and work for at least six months was a unifier that ensured the two countries have a platform for exchange and collaboration.

3.3.1 Partnership and Know-How Transfer and Training (KHTT)

While space promotes national prestige among the committee of nations, it also presents arena for partnerships and collaboration at various level with other countries and institutions. It assists to foster collaboration with partner countries at bilateral and multilateral levels.

The know-how transfer and training (KHTT) for instance ensures that participants in the manufacturing of any space hardware benefit as individuals and are also able to transfer the knowledge gained through the process to colleagues or students back home. Examples are the building of NigeriaSat-X with the help of Surrey Satellite Technology Ltd (SSTL) and most recently the construction of the Cube-Sats Nigeria EduSat-1 and GhanaSat-1 which were built in collaboration with partners from Kyushu Institute of Technology Japan through the Joint Global Multi-National Birds project as well as 1KUNS-PF, Kenya's first satellite which was designed and built with assistance from Japanese Aerospace Exploration Agency (JAXA) and the United Nations Office for Outer Space Affairs (UNOOSA).

So many other opportunities for partnerships in space are opening up, recently the Chinese National Space Agency (CNSA) signed an agreement with the United Nations to provide opportunities for astronauts from UN member developing countries to participate in China's future space stations and also carry out science experiments on board[19]. The initiative has the following aims:

- *promote international cooperation in human space flight and activities related to space exploration;*
- *provide flight experiment and space application opportunities on-board the CSS for United Nations Member States;*
- *promote capacity-building activities by making use of human space flight technologies, including facilities and resources from China's human spaceflight programme;*
- *and promote increased awareness among United Nations Member States of the benefits of utilizing human space technology and its applications.*

3.3.2 The UN/Dream Chaser Programme

The United Nations Office for Outer Space Affairs (UNOOSA) is currently in partnership with the Sierra Nevada Corporation (SNC)—now Northrop Grumman

---

[19]UNOOSA, 'United Nations and China Invite Applications to Conduct Experiments On-Board China's Space Station' (2018) <http://www.unoosa.org/oosa/en/informationfor/media/2018-unis-os-496.html> accessed 9 August 2018.

—to provide Member States with the opportunity to carry experiments, payloads, or satellites provided by institutions in the participating countries in an orbital space mission utilizing SNC's Dream Chaser space vehicle.[20] This is part of the UNOOSA Human Space Technology Initiative (HSTI) which aims to promote international cooperation, conduct outreach activities and support capacity building efforts worldwide. Developing countries participating in the UN/Dream Chaser mission will to build and fly payloads for microgravity science, remote earth sensing, hardware qualification and assets deployment.[21]

3.4 New Pathways to Space for Africa

With the advent of commercial space, there is opportunity for African countries to participate in the space sector at various levels without getting involved in expensive space programs and also provides the opportunity of partnering with international organizations and countries to obtain access to space. This allows for experts in universities and government agencies to execute space projects with limited human resources thereby facilitating and maximizing technology transfer while getting involved in the entire life cycle of the satellite.[22]

Possible approaches to leap frog into the space sector have been suggested by Karan Jani in two classes of a case studies[23]:

- *Case-A*: *a developing nation which uses international cooperation to advance its space-science program*
- *Case-B*: *a set of nations that collaborate to enhance a common space-science program or objective*

Above case studies are viable models that can be explored by prospective space countries as possible ways of getting into the space industry with little resources.

---

[20]UNOOSA, 'Orbital Space Mission' (2017) <http://www.unoosa.org/oosa/en/ourwork/psa/hsti/FreeFlyer_Orbital_Mission.html> accessed 17 July 2018.

[21]Luciano Saccani, 'United Nations Dream Chaser ® Mission - United Nations/Austria Symposium "Access to Space : Holistic Capacity Building for the 21st Century"' <http://www.unoosa.org/documents/pdf/psa/activities/2017/GrazSymposium/presentations/Monday/Presentation12.pdf>.

[22]Mazlan Othman, 'Small Satellites for the Benefit of Developing Countries' (2003) 52 Acta Astronautica 687 <https://www.dlr.de/iaa.symp/en/Portaldata/49/Resources/dokumente/archiv3/0102.pdf>.

[23]Karan Jani, 'Impact of International Cooperation for Sustaining Space-Science Programs' <https://arxiv.org/abs/1610.08618>.

4 Space Partnership and Benefits in Africa

This section of the document looks at the various structures and mechanisms set up to promote space effort on the African continent. It also explores different space related collaborative efforts and how they have contributed to furtherance of space activities among African countries. Possible approach with regards to adoption of new technologies, space access options including their associated benefits is also discussed.

4.1 Mechanisms for Promoting Space Partnership in Africa

The African Union (AU) having realized the importance of the use of space technology for strategic and developmental reasons has proposed the African Space Agency. The initial document has identified various areas space assets can contribute to the development of the African continent. The African Space Policy and Strategy that were adopted by the African Union Heads of State and Government during their Twenty-Sixth Ordinary Session on 31 January 2016 in Addis Ababa was conceived as the initial step towards realizing a viable African Space Programme.[24] Similarly, Agenda 2063 which is a strategic framework for the socio-economic transformation of the African continent was adopted by the Heads of State and Governments of the African Union at their 24th Ordinary Assembly held in Addis Ababa, Ethiopia, from 30–31 January 2015 with the goal of accelerating the implementation of past and existing continental initiatives for growth and sustainable development during the next 50 years after it was adopted.[25,26] The synergy between both adoptions will position space as a leading player in achieving the sustainable development goals in Africa.

One of the prominent demonstrations of partnership for space in the African continent is the push for an African Space Agency. The AU's African *Space Strategy* has listed the objective of this strategy. The description of some of the key challenges to societal needs in Africa is highlighted in the policy frame work of the provisional document shown in Table 2.

The *"indicative information & products"* section from Table 2 lists parameters that can be explored or monitored using space assets—ranging from Earth observation satellites to communication satellites—with potential to contribute to solving environmental and economic problems which forms the heart of the SDGs within the African continent.

---

[24]African Union, 'African Space Strategy' (2017) 16 <https://au.int/sites/default/files/newsevents/workingdocuments/33178-wd-african_space_strategy_-_st20445_e_original.pdf>.
[25]African Union, 'What Is Agenda 2063?' (*Agenda 2063*) <https://au.int/en/agenda2063> accessed 30 June 2018.
[26]African Union Commission, 'Agenda 2063' [2015] Final Edition 1 <http://www.un.org/en/africa/osaa/pdf/au/agenda2063.pdf>.

Table 2 A description of policy frameworks and corresponding response to key challenges on the African continent [African Union (n 24)]

Societal needs	Policy framework	Indicative information and products
Food security	Comprehensive Africa Agriculture Development Programme	Rainfall, yield, production, crops distribution, soil, land suitability
Water resources	African Water Vision 20125	Hydrography, aquifers, water bodies, quality, waste water and use
Marine and coastal zones	2050 Africa's Integrated Maritime Strategy	Coastal zones degradation, fisheries potential
Environment	NEPAD –Environment Action Plan (EAP)	Ecosystems, biodiversity, vegetation, land cover
Weather and climate	Climate Development Africa Integrated African Strategy on Meteorology	Rainfall, temperature, wind, aerosol, climate trends and extremes
Security and emergency	Africa Regional Strategy on Disaster Convention on Cyber Security and Personal Data Protection	Vulnerability, risk
Health planning	Africa Health Strategy	Disease vectors, environmental factors, population distribution
Governance and commerce	e-Government Strategy	Location- based mobile services, mapping of Government ICT infrastructures
Infrastructure	Programme on Infrastructure Development (PIDA)	Spatial information on key infrastructure, such as transport infrastructure, energy sources and power systems and distribution networks
Information and communications	Reference Framework for Harmonization of Telecommunications and ICT Polices & Regulation in Africa African Regional Action Plan for Knowledge Economy (ARAPKE)	Telecommunications, internet, TV broadcasting, mobile communications, e-Commerce, e-Government, e-Learning
Innovation	Science, Technology and Innovation Strategy for Africa	Food security, disease prevention, communications, security

4.2 Collaborative Space Programmes and Platforms in Africa

African Resource Management Constellation (ARMC): A solely African initiative, the African Resource Management Constellation (ARMC) which was formed to provide easy access for users in the fields of disaster management, food security, public health, infrastructure, land use, and water resource management. It was initially conceived around 2004 as the African Resource and Environmental Management Satellite Constellation with the goal of developing a constellation of satellites capable of providing real time, unrestricted and affordable access to satellite data for supporting effective environmental and resource management in Africa.[27] The ARMC agreement was signed in 2009 by four partner countries—Algeria, Kenya, Nigeria and South Africa.

Disaster Monitoring Constellation (DMC): Though few African countries have functional satellites in orbit there are levels of collaboration going on involving few African countries as well as some international partners. A very good example is the Disaster Monitoring Constellation (DMC) an international program proposed and led Surrey Satellite Technology Ltd (SSTL), UK for construction of a network of five LEO microsatellites capable of daily global imaging capability at medium resolution for rapid-response disaster monitoring and mitigation.[28] Table 3 shows countries that have been partners in the Disaster Monitoring Constellation—two African countries are represented—Algeria and Nigeria.

4.3 Benefits of Africa Space Programs

Participation in the space sector will boost African technology base and promote development thereby reducing poverty within the continent in so many ways. Some potential terrestrial application areas of space assets include communications, education, disaster management, education, agriculture, security, environmental protection and natural resource management.

The benefit and spin-off from space technology—both acquired and those that stem from KHTT would impact continental economy positively and tremendously contribute to promotion and interest in science, technology engineering and math (STEM) related educational programmes from primary to tertiary levels.

Satellites are launched based on objectives and requirements—driven by institutional or commercial motivations. A developing country for example would have series of reasons for short and long term goals motivation as described in the Table 4.

4.4 Possible Future Directions of Space Efforts in Africa

The current focus of most African countries that have launched satellites to space is directed towards space applications also most of the existing satellites procured and launched by few African countries are usually from the same set of manufacturers and have similar features but it is of importance to grow indigenous space technologies that will adequately target the challenges within the continent. This means that Africa would need to develop space technologies specifically tailored to the

---

[27]Simon Adebola, 'African Resource Management Satellite (ARMC) Constellation' (*I Initiative*, 2009) <https://iinitiative.wordpress.com/2009/12/21/african-resource-management-satellite/> accessed 19 July 2018.

[28]ESA, 'DMC-1G (Disaster Monitoring Constellation - First Generation)' (*eoPortal Directory*, 2018) <https://directory.eoportal.org/web/eoportal/satellite-missions/content/-/article/dmc> accessed 22 June 2018.

Table 3 DMC missions and partner organizations (ibid.)

Spacecraft/Mission	Country/Organization	Platform	Launch date	GSD	Swath width	Mission status
Alsat-1	CNTS (Algeria)	SSTL-100	2002	32 m MS	650 km	Completed
BilSat	Tubitak (Turkey)	SSTL-100	2003	32 m MS	650 km	Completed
NigeriaSat-1	NASRDA (Nigeria)	SSTL-100	2003	32 m MS	650 km	Completed
UK-DMC-1		SSTL-100	2003	32 m MS	650 km	Completed
Beijing-1[a]		SSTL-150	2005	32 m MS	650 km	Operational
UK-DMC-2		SSTL-100	2009	22 m MS	650 km	Operational
Deimos-1	Deimos Imaging (Spain)	SSTL-100	2009	22 m MS	650 km	Operational
NigeriaSat-2	NASRDA (Nigeria)	SSTL-300	2011	2.5 m PAN 5 m MS 32 m MS	20 km 20 km 320 km	Operational
NigeriaSat-X	NASRDA (Nigeria)	SSTL-100	2011	22 m MS	650 km	Operational
DMC-3 (a/b/c)	DMCii, with 100% capacity leased to 21AT of China	SSTL-300S1	2015	1 m PAN 4 m MS	23 km 23 km	Operational
NovaSAR-S	UKSA	SSTL-300 avionics	2015	6– 30 m SAR	15– 750 km	In development

[a]SSTL considers Beijing-1 as its first second-generation DMC satellite - due to an enhanced type platform (SSTL-150) of TopSat heritage

needs of each country or region. Some thematic areas that space can be effectively utilized are shown in Table 5.

While this is beneficial there is also the need to get to a stage where African countries will start to participate in space exploration programmes. This will involve active collaboration with partners carrying out exploration programmes of the solar system and beyond.

4.5 Challenge and Lure of Space

Space technology is usually a closely guided technology because of the dual use (military and civilian). Also, the cutting edge nature of space programmes allows the owner of such technology to be self-sufficient and utilize technologies that

Table 4 Framework for potential motivations for rational investment in satellite service, hardware, expertise and infrastructure within a developing country (Danielle Wood and Annalisa Weigel, 'Building Technological Capability within Satellite Programs in Developing Countries' (2011) 69 Acta Astronautica 1110)

Investment Area	Satellite Service: *Using satellite services in earth observation, communication, navigation and science*	Satellite Hardware: *Owning and operating a spacecraft and supporting ground system*	Expertise: Satellite *Training personnel in satellite engineering*	Satellite Infrastructure: *Establishing local facilities to fabricate satellites*
Short term motivation	Address time sensitive national needs for information	• Meet unique local requirements for information with specific temporal frequency, spatial resolution, spectral in satellite coverage	Develop knowledge to be an informed consumer of satellite services	Increase technical involvement of local personnel activities
Long term motivation	Enable informed regional planning Enhance infrastructure and industry	Gain operations experience • Decrease dependence on uncertain technology sources Ensure service continuity	Inspire young scholars • Enhance education and research opportunities Build industrial capability	Use infrastructure to facilitate long term series of satellite projects

allows for global competitiveness without the compulsion to use or adapt to controlled and expensive foreign technology. Space technology is a high end venture that requires multi-disciplinary contribution and permits the space sector to infuse diversity through collaboration from both public and private organizations thereby creating jobs and exposing countries to cutting edge technology. In recent times several African countries have woken up to the need to participate in space activities thanks to the spiral drop in price of space component.

Starting and maintaining a space program is capital intensive and requires human investment in high end technologies, also essential is the need to have the ancillary support systems needed to run them. This would be demanding for most countries in Africa due to paucity of funds.

There are schools of thought with regards to participation in space activities by low income countries, one is that such countries should rather invest their meagre resources into programmes that would be beneficial to the population - in short term, while another school of thought advocates for participation with the expectation that the long term benefit of investment into space programmes would surpass the former.

Presently the budgetary allocations earmarked for technology related programs in Africa are low, therefore going for expensive spacecraft would deplete already meagre budgets. New approaches that would permit entry to space by low income

Table 5 User needs mapped against the various space thematic areas [African Union (n 29)]

User needs	Earth observation											Navigation and positioning	Satellite communications	Space Science
	Spatial resolution								Temporal resolution					
	<50 cm	50 cm–1 m	1 m–2.5 m	2.5 m–5 m	5 m–10 m	10 m–20 m	20 m–30 m	>30 m	Daily	Seasonal	Annual			
Disasters	✓	✓	✓	✓	✓	✓	✓	✓	✓			✓	✓	✓
Health					✓	✓				✓		✓	✓	
Energy					✓	✓					✓	✓	✓	✓
Climate					✓	✓			✓			✓		✓
Water	✓	✓	✓	✓	✓	✓	✓	✓	✓			✓	✓	✓
Weather	✓	✓	✓	✓	✓	✓	✓	✓	✓	✓		✓	✓	
Ecosystems				✓	✓	✓	✓	✓		✓		✓	✓	
Agriculture				✓	✓	✓	✓	✓	✓			✓		
Biodiversity				✓	✓	✓	✓	✓			✓	✓		
Peace, safety and security	✓	✓	✓	✓				✓	✓			✓	✓	✓
Human migration and settlements	✓	✓	✓								✓	✓	✓	
Education and human resources				✓	✓	✓	✓	✓			✓	✓	✓	✓
Communications												✓	✓	
Trade and industry	✓	✓	✓	✓	✓	✓	✓	✓		✓		✓	✓	✓
Transport	✓	✓	✓	✓	✓	✓	✓	✓			✓	✓	✓	
Infrastructure					✓	✓			✓			✓	✓	

economies are becoming available through partnerships and cheaper commercial space entities thereby offering so many options to choose from.

4.6   Small Satellites/CubeSats to the Rescue

Two important disruptions are advantageous to aspiring space nations in Africa - the access to small satellites and the advent of commercial space market. These combinations provide opportunity for African countries to participate in the space sector at various levels without out necessarily engaging in expensive and costly space programs. In the past only countries with huge capital could launch satellites into space but the advent of small satellites and even smaller ones in the form of CubeSats has enabled several African countries to aspire and belong to the group of countries that have 'satellites in space'. Small satellites most especially CubeSats have paved way for more countries to be part of space programmes that was previously reserved only for rich countries and some African countries have towed this line by investing and ultimately launching CubeSats. Countries such as Kenya and Ghana were able to loft their CubeSats to space. This is a milestone towards playing a role as a participant in the space arena.

Capacity building in the area of space technology for developing countries is benefiting from small satellites that emanated from rapid development of and the use of commercial off-the-shelf technologies that are cheaper as well as advanced. This provides that platform to focus on specific missions that can reduce the time spent on the development process while providing faster and cheaper access to space.[29] It also affords the possibility of achieving comparable results to very expensive satellites/space systems—with only a downside of mission lifespan and limited payloads (Table 6).

The entry of CubeSats has drastically reduced the cost of manufacturing "spacecraft" when compared to the cost of big satellites such as Landsat which, according the a 1983 report cost approximately $573.1 million[30] while the next addition to the Landsat series (Landsat 9) is expected to cost about $650 million.[31,32] The access to space experience could provide impetus for more investment into research and development (R&D) in Africa. African R&D has witnessed a growth of 0.88, 0.90 and 0.92% in 2016, 2017 and 2018 respectively from the world total. African R&D investment has increased from $18.91 billion in 2017 to

[29]Othman (n 22).

[30]US Government Accountability Office, 'Costs and Uses of Remote Sensing Satellites' (1983) <https://www.gao.gov/products/120782> accessed 23 June 2018.

[31]Dan Leone, 'NASA Official: A Landsat 8 Clone Would Cost More Than $650 Million' (*SpaceNews*, 2014) <https://spacenews.com/40841nasa-official-a-landsat-8-clone-would-cost-more-than-650-million/> accessed 23 June 2018.

[32]Peter Folger, 'Landsat: Overview and Issues for Congress' 15 <https://fas.org/sgp/crs/misc/R40594.pdf>.

Table 6 Classification of artificial satellites

Satellite classification	Mass (kg)
Large satellite	>1000
Medium satellite	500–1000
Mini satellite	100–500
Micro satellite	10–100
Nano satellite	1–10
Pico satellite	0.1–1
Femto satellite	<0.1

Gottfried Konecny, 'Small Satellites—A Tool for Earth Observation?', *International Society for Photogrammetry and Remote Sensing 20th, International congress for photogrammetry and remote sensing; ISPRS XXth congress* (Organising Committee of the XXth international congress for photogrammetry and remote sensing 2004)

$20.07 billion in 2018[33] which could positively impact the space sector and contribute to poverty reduction. Not only will space position the developing countries in Africa to play a role in the international community it will also empower the populace to benefit and also make contribution towards emancipation from poverty and achieving the SDGs on the African continent.

5 Conclusion

This chapter has briefly chronicled the emergence of active space from passive space in Africa, it also explored few of the African space heritage and current trend in the space sector within the continent.

Space technology and assets are poised to contribute to the narrative of sustainable development especially in the area of poverty reduction - critical elements in the SDGs - and also promote prosperity and partnership. Space is helping to infuse inclusiveness by allowing collaboration from both public and private organizations thereby creating jobs and exposing the host countries to cutting edge technologies. Various strategies and policies designed to achieve this in conjunction with space technology were also touched on.

Space assets are strategic to the achievement of the SDGs as they affect almost every facet of life in the modern era and could contribute tremendously to economic development in developing countries thereby helping reduce the gap between poverty and prosperity within the African continent.

---

[33]R&D Magazine, '2018 Global R&D Funding Forecast' <https://digital.rdmag.com/researchanddevelopment/february_2018>.

Author Biography

Samuel Anih developed a profound interest in space exploration during his high school days and later founded SpaceRovers which acted as a melting pot for fellow space enthusiasts as an undergrad at Obafemi Awolowo University, Nigeria. He received an MSc from the International Space University (ISU), Illkirch-Graffenstaden (Strasbourg), France and later a graduate fellowship at NASA Ames, Moffet Field, California. He has more than 10 years working experience as a scientific officer at the African Regional Centre for Space Science and Technology Education in English (ARCSSTE-E), a United Nations affiliated centre. He is currently a PhD candidate at SpaceLab in the University of Cape Town. Sammy is passionate about space science & technology, most especially the future of crewed space missions beyond low Earth orbit, sustainability, planetary protection and space safety. He hopes to help build a new generation of space explorers and leaders.

Sub-Saharan Africa: "Info-Agritech" a Potential Game Changer

Christoffel Kotze

Abstract

The UN 2030 SDG (Sustainable Development Goals) identifies agriculture as one of the sectors unparalleled in terms of its potential to support human development whilst, simultaneously, supporting sustainable economic growth. By assisting farmers in developing countries to become as least as productive as their counterparts in the developed world, agricultural productivity can therefore be used as a key strategic socio-economic change agent. In Africa and in particular sub-Saharan Africa, agriculture has a significant socio-economic impact. This sector is on average the largest source of employment, but plagued by low productivity, it does not translate into an equal quantum of economic contribution. Sub-Saharan Africa in particular is well suited to benefit in this space, with a rapidly growing population, large under-utilised agricultural areas and low traditional agricultural productivity. Rapid simultaneous technological development in the so-called SMACT (BTC http://www.boston-technology.com/smact/ accessed 5 July 2018) group in particular holds great promise to assist farmers in developing world. The idea of this essay is to explore how satellite technology can act as an enabler of "Info-Agritech"—(the application of information technology in agriculture)—in terms of the sub-Saharan Africa agricultural sector.

Keywords

Agriculture · Digital divide · Info-Agritech · Last-mile · Satellite technology · Sustainable development goals

C. Kotze (✉)
University of Cape Town, Cape Town, South Africa
e-mail: ktzchr005@myuct.ac.za

© Springer Nature Switzerland AG 2019
A. Froehlich (ed.), *Embedding Space in African Society*, Southern Space Studies,
https://doi.org/10.1007/978-3-030-06040-4_3

1   Introduction

"Technology has the potential to answer some of our biggest questions and help us better understand the world around us. In almost every industry, massive efforts are underway to connect our physical and digital worlds, unleashing the Fourth Industrial Revolution. We must work together to ensure that the food and agriculture sector is not left behind—and that these efforts contribute towards global food systems that benefit farmers, consumers and the planet."—Bernard Meyerson Chief Innovation Officer, IBM Corporation.[1]

Though the world does produce enough food to feed all its citizens, hunger is still an unfortunate reality for a large portion of the global population. This is present in all countries to a certain extent, but is much more prevalent in poorer countries. Almost 98% of world hunger is associated with underdeveloped countries, where on the upper end of the scale, 73% of the population could be affected.[2] Malnutrition and chronic hunger has a profound effect on the people it afflicts. "Food-insecurity"–(a situation where there is uncertainty in the ability of a country or an individual to have access to adequate food supply)—is a primary contributing factor. This condition causes stress in individuals and communities alike and can— lead to strive and political instability. The opposite of this scenario is —"food-security", defined by the World Food Summit as creating a condition where "all people at all times have access to sufficient safe and nutritious food for an active and healthy life."[3] Hunger can be caused by a number of diverse factors, but persistent hunger is primarily caused by poverty, with the—poor lacking the resources to buy food directly or indirectly, i.e. not having access to the means of production. Malnutrition is a condition—especially pertinent to infants and developing children—(where people do not have access to essential nutrients)—which has a profound long-term socio-economic impact. Adequate trace elements and micronutrients are crucial for normal, healthy development and growth in humans. In developing countries, three of these are of particularly importance according to the WHO[4] (World Health Organisation) Iron, Vitamin A and Iodine, with deficiencies creating a severe long-term impact on an individual's growth and development. A person affected by chronic malnutrition will therefore be less able to compete in the labour market, causing poverty, leading to a vicious circle making it difficult for the affected person to escape.

The UN 2030 SDG[5] (Sustainable Development Goals) identifies agriculture as the economic sector linking—all of the 17 UN 2030 SDG's in one way or the other, as it addresses a foundation issue related to the human condition, namely the

---

[1]Bernard Meyerson Chief Innovation Officer IBM as WEF agenda contributor.
[2]The Borgen Project https://borgenproject.org/ accessed 7 July 2018.
[3]COMMITTEE ON WORLD FOOD SECURITY—Thirty-ninth Session, 2012 http://www.fao.org/bodies/cfs/cfs39/en/ accessed 8 July 2018.
[4]WHO http://www.who.int/elena/titles/full_recommendations/sam_management/en/ accessed 7 July 2018.
[5]'The Sustainable Development Agenda' www.un.org/sustainabledevelopment/development-agenda accessed 15 January 2018.

production of food. Agriculture is therefore intrinsically linked to the use of land, water and energy. It is estimated the sector consumes an estimated 70% of all fresh water—and 30% of all energy which is mostly fossil fuel based. The industry, for obvious reasons, is also exposed more than any other industry to the effects of climate change to which it ironically indirectly contributes to. Food systems are deemed to be responsible for up to 30% of global GHG production. Deforestation in 2016 was responsible for the loss of 30 million tree covered hectares,—(50% more than in 2015)—of which agriculture was the primary driver.

By assisting farmers, especially small-scale farmers in developing areas such as sub-Saharan Africa, to embrace sustainable production and consumption, all SDG's will benefit. It is here where technology can play a crucial role.

2 Agriculture in Sub-Saharan Africa

Agriculture is the single largest employer in the world. It is undoubtedly also the cornerstone of any developing economy through the provision of food security, job creation, social welfare and as a contributor to the GDP. Ensuring a healthy agricultural sector is also a key contributor to political stability. Hendrix[6] and Brinkman —explored the link between food insecurity and violent conflict. It appears that higher food prices—(normally the result of dependence on food imports)—can lead to civil conflict, protest, rioting and violent conflict between communities, even total democratic breakdown. The spark of the political revolutions in Egypt and Tunisia in 2011 was amongst others linked to record high food prices.[7]

Ironically, despite the availability of vast agricultural resources in the form of large areas of arable land, Africa is the continent with the highest incidence of undernourishment worldwide (affecting around 25% of the population). Due to low levels of agricultural productivity and post-harvest losses of around 30% of average production, Africa is forced to import food. Annually an estimated US $25 billion worth of staple is imported, money which could be used better elsewhere. According to data[8] from the World Bank in 2017, agriculture represented 57% of all employment in sub-Saharan Africa (with 47% of the labour done by woman) as opposed to OECD member countries which was just under 5%. The contribution of agriculture to GDP on the other hand is significantly lower with the 2017 figure for sub-Saharan Africa at just over 16%. The relatively low GDP contribution of agriculture to the economy, when taking into account the high employment the sector provides, is a sign of low productivity and limited value-add to agricultural commodities. This sector represents the main source of income for the rural population, estimated to be as high as 64%. In sub-Saharan Africa, agriculture is dominated by the production of

---

[6]Brinkman, Henk-Jan, and Cullen S. Hendrix. "Food Insecurity and Violent Conflict: Causes." Consequences, and Addressing the Challenges (2011) World Food Programme.

[7]Lagi, Marco, Karla Bertrand, and Yaneer Bar-Yam. "The food crises and political instability in North Africa and the Middle East." (2011).

[8]The World Bank https://data.worldbank.org/indicator/NV.AGR.TOTL.ZS accessed 10 July 2018.

a variety of traditional staple food crops like (cassava and maize), supplemented by cash crops (coffee, cocoa, tobacco etc.) A very high percentage (80%) of agricultural land in sub-Saharan Africa takes the form of smallholder farmers using small plots with minimal additional inputs on which low-yielding crops are cultivated. These small farms are also heavily dependent on the weather, as rainwater is the primary water source for crops. Agricultural productivity in sub-Saharan Africa is considerably lower when compared to other developing regions, illustrated by the fact that cereal yields in sub-Saharan Africa have seen very little year on year increase from 1960, as compared to other developing economies which has seen a sharp increase over the same period.

One of the problems affecting the small-scale farmers is the lack of access to adequate agricultural advisory services. At the 2015 BEANSA[9] conference one of the key recommendations made by delegates reads "…… agroforestry information should be synthesized for easy comprehension by farmers, this will demand new ideas and innovation to enable the departure from the old ways of thinking". The aforementioned does not only apply to small-scale farmers—and even established framers can encounter the "information drought". At the 'Beating Famine'[10] Conference held in April 2015 in Malawi, the plight of access to information by the sub-Saharan Africa farmer was highlighted again; Information collected about various aspects of agriculture is collated and published as research papers, publications and technical manuals—yet it rarely reaches the ones who can benefit from it most—the farmer. In the event that such information does find its way to the farmer, it is more often than not in a very technical format and not in a local language understandable by the farmer. The importance of local language and customs, when it comes to the distribution of agricultural information is extremely important. When farmers do not have access to new relevant information they will stick to old methods. Methods which are not necessarily suitable to the changing environment—driven by climate change—around them.

The key to the Sub-Saharan Africa agricultural renaissance therefore lies very much in the ability to provide all farmers with pertinent, relevant, timeous information in a simple easy to understand format and a language they are comfortable with to ensure the message is brought across correctly. The solution is therefore apparent—simply bring the right information to the right people at the right time.

3   Technological Impact on Agriculture

According to the WEF's "Big Twelve", technology has the ability to have a profound impact on agriculture and in the process, to realize significant financial and environmental benefits. One of the most prevalent game changing technologies

---

[9]http://blog.worldagroforestry.org/index.php/2015/04/20/bleansa-the-makings-of-an-evergreen-agriculture-hub-for-southern-africa/ accessed 9 July 2018.
[10]http://beatingfamine.com/beating-famine-conference-declaration/.

accessible to even the poorer members of society are internet connected smart mobile devices i.e. smartphones or tablets. According to the—2018 WEF[11] study "Innovation with a Purpose: The role of technology innovation in accelerating food systems transformation"—if mobile technology alone was rolled out to at least 70% of all farms worldwide by 2030 it is estimated to effect the following:-

- Generate up to US $200 billion of income for farmers.
- Reduce equivalent GHG emissions by up to 100—mega tonnes of CO_2.
- Reduce freshwater withdrawals by up to 100 billion m^3.

Smartphones or internet connected devices like tablets have the ability to make a significant impact at a relative low cost to the farmer. It allows access to financial services and payment systems. Importantly, it can provide the day to day management information to assist the farmer at operational level, including weather forecasts. For emerging farmers it can provide insight into market and demand trends and links to supply-chain nodes. It is estimated that if 300 million farms adopt mobile-based services by 2030 an additional 250–500 million tonnes of food could be produced annually whilst preventing the loss of at least 20 million tonnes. This will only be achievable provided a minimum infrastructure and education is available and accessible by the all farmers including illiterate famers in unconnected areas.

This approach is characterised by the collection and processing of data to produce information systems which can be used to aid the farmer e.g.—creating advanced prediction models and presenting it in a plain useful way. Farming with the aid of long prediction systems helps the farmer to plan and mitigate risk e.g. weather and plague prediction systems. More than 80% of agricultural production in the developing world is produced by small-scale farmers who typically do not have access to agricultural support services. Plant diseases thus often are not identified early on in the infection cycle leading to devastating crop losses. A research project using artificial intelligence which can learn to recognize specific plant diseases very accurately, has been developed by researchers at Penn State and the Swiss Federal Institute of Technology. This technology could be particularly beneficial to small-scale farmers in areas such as sub-Saharan Africa. Mobile devices can be used to gather data, which can be fed to the system to prescribe timeous intervention. For the purpose of this essay, the collective term "Info-Agritech" will be used to refer to the application of information technology in agriculture. Communication infrastructure—is key for any of these services to be of any benefit, which is a problem for the global masses affected by the so-called "Digital Divide".

---

[11]https://www.weforum.org/reports/innovation-with-a-purpose-the-role-of-technology-innovation-in-accelerating-food-systems-transformation.

4   The—"Digital Divide"

The US department of Commerce defined the "Digital Divide" as the difference between two groups of technology users: on the one extreme, a group of people with access to the "latest and the greatest"—in digital tools and training as opposed to the other end of the social spectrum where a group exists with little to no access of any of the aforementioned "digital privilege". The reason for the existence of the "digital haves/have-nots" has been studied extensively and causative factors have been found to be a combination of social and spatial factors—e.g. where the person lives, income level, age, education, gender and even ethnicity (Fig. 1).

At a macro level, countries with high levels of digital penetration will find it increasingly difficult to trade with countries with low levels of digital integration at ground level. In sociology the "Matthew Effect"—(a term coined by Robert K. Merton[12]) refers to a situation where advantage propagates further advantage and vice versa—i.e. the rich get richer. The same can be said for nations, with those lagging in digital adoption typically are more economically disadvantaged and therefore not able to play catch-up, subsequently significantly hampering their future ability to compete economically, leading to further economic impairment in a negative spiral. The digital divide is a multi-faceted concept involving a number of digital technologies, however the availability of internet and bandwidth is—the key to the digital economy. Internet access is thus generally regarded as a metric to

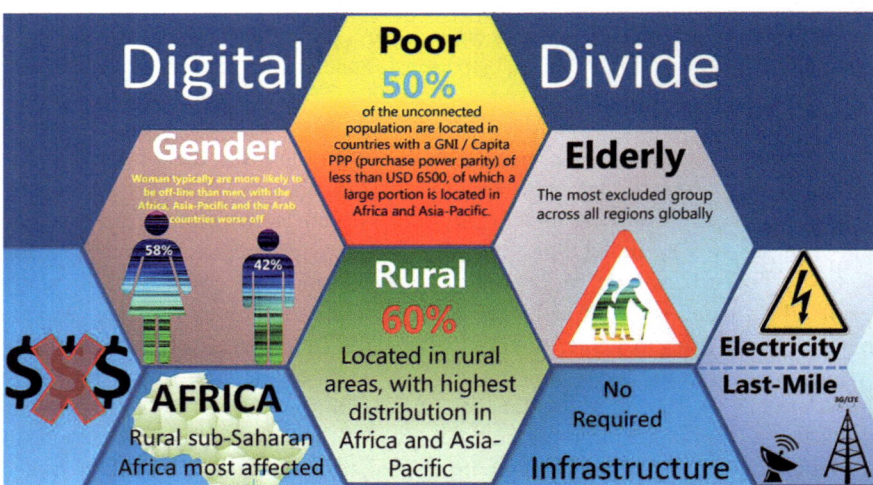

Fig. 1 The "Digital Divide" at a glance (ITU, 2017. ICT Facts and Figures 2017. Available at: https://www.itu.int/en/ITUD/Statistics/Documents/facts/ICTFactsFigures2017.pdf. Accessed 16 February 2018)

---

[12]Robert K. Merton "The Matthew effect in science: The reward and communication systems of science are considered.' (1968) Science 159.3810 p56.

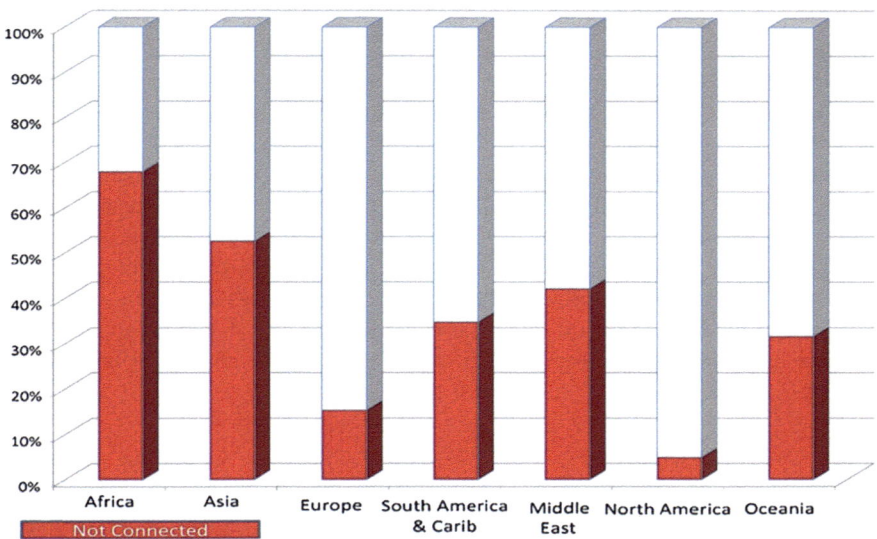

Fig. 2 People connected to the internet globally 2017 data (ITU, ICT Facts and Figures 2017 https://www.itu.int/en/ITU-D/Statistics/Pages/facts/default.aspx. accessed 14 July 2018)

gauge the presence and extent of digital divide for a specific demographic. The availability of broadband infrastructure is still the primary exclusion factor for— un-connected communities worldwide though. Africa has the lowest global rate with less than 35% of the total population connected as opposed to North America with more than 90% connected (Fig. 2).

Last-mile technology is a figurative term used in the telecommunications sector technology to describe the final connectivity leg between an internet service provider service provider and an individual user.

It has nothing to do with physical distance, it could be very short as in the link from a telephone pole to a house or thousands of miles in the case of a satellite link. For a large portion of the global population the provision of the last-mile remains the principal problem preventing the provision of broadband internet services. The last-mile challenge is the result of a number of factors with remoteness and lacks of infrastructure typically primary barriers to connection. The more remote the area, the higher the likelihood of no connection, as putting down physical cable infrastructure such as fibre optics is impractical or not even possible due to cost, accessibility and/or maintenance issues. There are a number of ways to connect the "Last-Mile" and they are normally deployed in a "fit-for-form" fashion depending on the nature of the last mile customer location guided by the 3A's: availability, accessibility and affordability. The ISP as a commercial enterprise will typically not deploy to an area if there is no financial incentive to do so. The World Economic

Forum[13] cites—lack of infrastructure to be the main reason why almost a third of the global population cannot connect, 31% with no 3G coverage and 15% without electricity including almost a third of Sub-Saharan Africa.

In addition to connectivity, bandwidth distribution per user is also unequal with an average of 140 kbps per user in the developed world versus only 6 kbps per user in the least developed countries. Available bandwidth is a key determinant of what type of content a user will be able to engage with effectively. Low bandwidth therefore generally creates a secondary level of inequality, preventing users from meaningfully interacting with feature rich content. A restricted bandwidth user, in a developing country will be prevented from meaningful access to video rich content and -peer-to-peer interactions characteristic of remote education applications. Satellite broadband technology can provide adequate bandwidth solutions not dependent on the "last mile" location of the user.

5   Satellite Technology

On October 4, 1957, the USSR placed the first man-made object into orbit in the form of an artificial satellite called "Sputnik". Whilst the initial use of space technology was largely driven by a political/military agenda inspired by "Cold War" zeitgeist, the commercial aspect soon became apparent. Space technology—is arguably one of the most powerful set of technology tools that can be applied to sustainable development. Currently there are more than 1700 working satellites orbiting the earth—with the ability to cover large geographic areas undeterred by remoteness, weather systems, conflict or extreme climatic conditions. Though the modern commercial satellite industry is dominated by communication satellites—representing 35% of all mission types—the impact of GPS and remote observation data is far reaching for the agriculture industry. Satellite technology in particular can enable the two stated requirements for "Info-Agritech" namely connectivity and data collection.

5.1   Satellite Broadband

Eutelsat KA-SAT 9A,[14] a very large HTS (High-Throughput Satellite) placed into GEO (Geostationary Equatorial Orbit) in late 2010, could be considered the first dedicated broadband satellite. Deploying spot beam technology, it started to provide high speed internet to users. Subsequently the broadband satellite industry has been undergoing significant technological innovation enabling development of HTS

---

[13]WEF 4 reasons 4 billion people are still offline https://www.weforum.org/agenda/2016/02/4-reasons-4-billion-people-are-still-offline/ accessed 7 June 2017.
[14]EutelSat https://www.eutelsat.com/en/satellites/the-fleet/EUTELSAT-KA-SAT.html accessed 1 July 2018.

in lower MEO (Medium Earth Orbit) and LEO (Low Earth Orbit) orbits. Currently the broadband segment is well poised for growth with at least five major systems available, Eutelsat, SES-O3b, ViaSat to name a few. In addition to the expansion of existing systems there have been a number of "mega constellations" announced[15] by amongst others Boeing, OneWeb and SpaceX. Frequency applications lodged by the three companies at the FCC indicate a total of almost 16 000 broadband satellites if all were to be deployed. To put this figure in perspective according to the current UCS[16] (Union of Concerned Scientists) satellite database by 31 August 2017 there were 1738 operational satellites in orbit of which 742 were communication satellites. These systems will add significant capacity to not only satisfy increased commercial demand but importantly provide a global footprint to deliver internet to the unconnected. Initiatives such as ITU's "Connect 2020" campaign—to bring 60% of the world's population online by 2020—of which a key barrier is connection of the large rural, largely poor population will benefit from these systems. Satellite broadband technology is particularly well-suited for large scale broadband internet delivery to rural and other underserviced areas and has as such been included in the NBP's (National Broadband Plans) of many governments. Though there are a number of—"Commercial Off-The-Shelf" systems available, suitable for general use and for particular use environments e.g. for a yacht or a caravan utilising parabolic dishes and solar power systems already integrated into a structure, they are generally not targeted at the poor rural communities where it can play a major role by linking the community to the internet.

5.2 Data Collection

Data collection and analytics are crucial to Info-Agritech systems in one way or the other, fortunately constantly evolving techniques allow for the fast analysis of data from different sources (terrestrial and satellite) to produce actionable information. This type of information can then be supplied to the farmer or any other stakeholder to give insights on the current crops and or build a risk profile on the crop. Orbiting satellites provide an unmatched opportunity to collect unique datasets such as—multispectral Earth imagery and location meta-data, not possible in other ways from any area on earth at any time. In conjunction with ICT developments enabling rapid storage and processing of large datasets, remote location and imagery data can be used to create ever more advanced GIS (geographic information system). An example of this approach in practice is demonstrated by—an ESA incubation program—start-up company in Germany namely Greenspin.[17] Merging current and historical data using analytical techniques including "deep learning" neural nets and

---

[15]Samantha Masunaga, "Satellite constellations could be poised to challenge the broadband industry" (2017) https://phys.org/news/2017-01-satellite-constellations-poised-broadband-industry. html accessed 10 July 2018.

[16]UCS Satellite Database https://www.ucsusa.org/nuclear-weapons/space-weapons/satellite-database#. W02-eLh9jIU accessed 1 July 2018.

[17]Greenspin https://www.greenspin.de/ accessed 5 July 2018.

machine learning allows it to produce unique new insights regarding crop conditions. The crop classification and quantification information is presented to the user via an easy to use mobile device or desktop application for use by any number of stakeholders including, commodity brokers, crop insurance etc.

6 Info-Agritech

Prediction has always played an important role in farming. The further the farmer can peek into the future the greater the opportunity to mitigate risk. In addition to the well-known weather prediction systems—both long and short term—market and plague prediction systems are very important to farmers. Though there are already many systems available, most of it require the use of propriety technology and integration techniques not within reach of many small-scale rural farmers —even if the communication technology was available. The Info-Agritech approach aims to remove the complexity out of these diverse collections of information and prediction models and present the farmer with a practical easy to use interface.

Plagues and—insect swarms like armyworms, aphids and weevil can destroy vast crop areas annually. Proactive pre-swarm forming prediction models can prevent the—problem and therefore it can be solved before it starts. The Food and Agriculture Organization of the United Nations (FAO) has devised a locust prediction system. This system allows for a 70 day prediction window using satellite data. Using combined data from different satellites,—NASA's Aqua & Terra and SMOS from ESA,—it creates risk maps highlighting areas which might favour locust swarming conditions.[18] The system warns small-scale farmers via a connected internet device such as a cheap Android tablet, if their area is potentially at risk. Similar systems can be deployed based on intelligent software analysis of satellite spectral data for plant disease prediction, soil management, crop stress detection etc.

In the Philippines more than 40 million coconut trees have been lost by farmers since 2013—primarily due to weather events and pests. A coconut tree takes up to twenty years to become fully productive. For small-scale farmers the loss of substantial number of productive trees can lead to financial ruin. The Grameen Foundation led partnership launched "FarmerLink[19]" in 2015 to assist small-scale farmers technologically. This "first-of-its kind" system works by providing the farmer with advisory services and training to increase sustainable practices and productivity. The system also provides the farmer with proactive warning systems to their mobile devices about severe weather events and the early detection of pests and plagues. These early warning systems allow for timeous intervention by

---

[18]ESA   https://www.esa.int/Our_Activities/Observing_the_Earth/Satellites_forewarn_of_locust_plagues accessed 28 May 2018.
[19]https://grameenfoundation.org/tags/farmerlink accessed 9 July 2018.

implementation of preventative strategies. Importantly the system also serves as a link to the market place and financial services. The heart of the system is the effective combination of data collected by satellite and field agents using connected mobile devices.

In Bangladesh, a country where most farming is done manually, a system has been designed to present complicated integrated information to farmers in such a way that it can easily be interpreted.[20] This Android based system can use either a cheap tablet or smartphone, which the farmer use to select the crop, presented by the system, with the best potential for the prevailing conditions from a number of options. An algorithm integrates predictor variables which is then presented to the farmer as a cultivation schedule and required daily inputs i.e. irrigation times, adjuvants and fertilizers.

In Zambia social media has been deployed as Agri-Tech to assist small-scale farmers. This system is essentially a Facebook group called "Small Scale Farmers Community[21]". The more than 6500 members, use it to share market information, ideas and trends at no cost.

RIICE[22]—(Remote Sensing-Based Information and Insurance for Crops in Emerging Economies) is a public-private funded system which uses ICT to collect information about rice growth in Cambodia, India, Indonesia, Philippines, Thailand, and Vietnam aiming to increase the rice crop. In addition to traditional terrestrial ICT, space based technology plays a central role in the system. The system works by;

- Data collection—Using ESA (European Space Agency) radar equipped remote sensing satellites to collect data about the Southeast Asia earth surface to detect even subtle changes.
- Mapping & Forecasting—Bespoke analysis software interpret the collected data into useful maps representing a number of indicators relating to rice growing in Southeast Asia e.g. Actual rice harvest yield information, In-season rice yield forecasts etc.
- Reducing Farmer Vulnerability—The rice yield observation data and yield forecasts can be used to proactively construct an intervention strategy for areas predicted to suffer crop losses e.g. famine relief organisations can plan accordingly and insurance companies can use the data to monitor the extent of crop losses.

Reuters Market Light[23] is another example of a decision-support system for farmers using a mobile platform. It provides "personalised" data analytics, fed by a team of full time agricultural advisors the system provides "relevant, timely, reliable

[20]Siddique, Talha, et al. "Automated farming prediction." (2017) Intelligent Systems Conference (IntelliSys), 2017. IEEE.
[21]SmallScaleFarmers http://www.agritech-expo.com/SmallScaleFarmers-PR accessed 2 July 2018.
[22]RIICE http://www.riice.org/ accessed 2 July 2018.
[23]Shamita http://www.samhita.org/social-organisation/reuters-market-light-rml/ accessed 7 July 2018.

and accurate information"—on more than 450 crops to more than 1.3 million registered farmers across India. Farmers can choose to receive information via internet applications, SMS or even voice in their preferred local language.

Mobile devices in conjunction with satellite data can now also provide small-scale farmers, who previously did not have access to the benefit of crop insurance products. A South African start-up company is such an example, "Mobbisurance[24]" facilitates crop insurance for smallholder farmers based on data provided via the South African National Space Agency from ESA/NASA satellites. The company provides crop monitoring services to its customers which are communicated via mobile phone.

As seen from the examples discussed in this section, there are many systems on the market potentially suitable to both existing and new small-scale farmers. Info-Agritech presents vast potential especially in Sub-Saharan Africa where it is being adopted at a very rapid rate, showing a 110% growth in start-ups since 2016 already. The expansion of digital infrastructure however will be essential as all the technology discussed has one thing in common—a connection to the internet to provide the essential networking required by the data flow. Therein lies the problem as currently almost a third of the population of the earth cannot connect to the internet, most of them poor people in the developing world, victims of the digital divide.

7 Opportunities in Sub-Saharan Africa

The key to achieving increased agricultural production—lies in the effective sharing of knowledge—over and above investment in the agricultural industry. Advisory information will empower all famers, but especially the emerging farmer, to embrace and utilise what the information economy has to offer. As the previous sections demonstrated mobile technology based systems are available with many examples successfully deployed worldwide. Though it might not be directly transportable from one area to the other, it does demonstrate the efficacy and potential of mobile technology. These ideas can be used and localised—to be relevant, reflecting local language and taking into account local practice and conditions, provided the investment is made to provide the basic infrastructure to bridge the digital divide. What makes the sub-Saharan Africa potential for an agricultural renaissance particularly good is that it's not constrained by legacy technology and therefore can embrace new trends and techniques, making it an ideal candidate to embrace new technology.

A particular challenge to overcome will be to supply "proof" to a sceptic target audience—where literacy might be a severe challenge, to illustrate the value of the long-term use of the product up-front. To "sell" the benefits of a system like this—even if all the relevant services are provided free of charge—will be challenging.

---

[24]Mobbisurance https://www.f6s.com/mobbisurance accessed 8 July 2018.

In areas where traditional ways of agriculture are very entrenched and even integrated in cultural practice reluctance to change might make implementation impractical. Essentially the traditional farmer will be required to forego traditional methods in favour of an unknown technology using foreign methods—the result of which will only be proven at the end of the crop cycle. This might be overcome by providing a proof of concept "crop section" to willing members of the community which can be used to prove the virtues of the new system. However, as cooperation from the user community will be essential in the form of user feedback, to grow and improve service to the community it aims to serves—it might be necessary to consider a different demographic, than existing small-scale farmers for the initial implementation.

Strategically, a "ready and willing" demographic, with some form of exposure to technology already might improve the implementation success rate considerably. To skilled young people who might be unemployed in the cities, introduction of this type of technology might serve as incentive to return to the land. Supplied with the relevant technology, they can apply their existing digital skills to—become not only high productivity "techno small-scale farmers", but also co-developers of such a system. Assisted by technology this new generation of farmers have the potential to positively impact the social-economic landscape of their country. They will also be able to assist in diffusion of the technology to the farmers steeped in traditional ways and conversely so incorporate effective ideas picked up from the traditional method into the system. A particular opportunity exist in countries such as South Africa, where land reform and restitution programs will create the opportunity for many people to enter the agricultural space.

8 Conclusion

Globally, agriculture is a powerful change agent with a significant impact on the overall success of the 2030 SDG's, though in sub-Saharan Africa it is currently not playing the full role that it can. Info-AgriTech can be a game changer, though in sub-Saharan Africa there are a number of challenges to overcome first. Availability of infrastructure in the form of internet connectivity (and electricity—to a certain extent) is a key requirement for these systems to function. Satellite technology can assist by bridging the "Last Mile" and in so doing closing the "Digital Divide"—the technology is also crucial to collect the data required for Info-Agritech.

Though agriculture as sector in Sub-Saharan Africa is the largest employer, it does not contribute equally to GDP due to productivity issues,—therein lies both the problem and the opportunity. Representing a strategic industry as far as employment is concerned, it can be a very powerful economic influencer provided the low productivity can be improved. Small-scale agriculture has relatively few barriers to entry, as opposed to other production industries such as mining for example—provided the land is available. In countries where land reform and restoration programs are active a special opportunity presents itself for agriculture.

By equipping beneficiaries of new land opting to go into agriculture with Info-Agritech, it will greatly improve their chance of success to engage in surplus agriculture production, leading to job creation, poverty alleviation and overall economic improvement.

Author Biography

Christoffel (Chris) Kotze, after a successful corporate career spanning two decades decided in 2012 to return to the Cape Town to establish a boutique strategy consultancy with a focus on ICT4SD. A technology strategist with a special interest in how technology can be used to promote sustainability—current research interests include the relationship between space technology, dematerialisation and the digital divide. Currently in process of completing a MPhil in Space Science at the University of Cape Town. Other qualifications include a BCom Honours (Information Systems)—UCT, BSc (Physiology & Microbiology)—UP, Diploma in Datametrics (Computer Science) UNISA, a number of executive courses at UCT Graduate School of Business. ISACA Certified in the Governance of Enterprise IT (CGEIT), TOGAF 9 Certified (Enterprise Architecture).

A Contribution to an Advisory Plan for Integrated Irrigation Water Management at Sidi Saad Dam System (Central Tunisia): From Research to Operational Support

W. Abdallah, M. Allani, R. Mezzi, R. Jlassi, A. Romdhane, F. Faidi, Z. Daouthi, A. Amara, H. Selmi, A. Zouabi, K. Selmi, T. Ayoub, R. Béji, M. A. Trabelssi, F. Joumade-Mansouri, E. Chalghaf, F. Stoffner, M. E. Hamza, H. W. Müller and A. Sahli

Abstract

In the agriculture sector, combining physically based soil water balance and simulation models with GIS (Geographic Information System) tools is of a considerable interest to manage the available water amount. Indeed, this combination can enhance water supply management, optimize agricultural catchments management and study impact of management intervention from small scale (plot) to a larger one, such as irrigated district and/or region. This work presents the case of Sidi Saad Dam System (central Tunisia). The main

H. W. Müller—Head of CREM-BGR Project.

W. Abdallah · M. Allani · R. Mezzi · R. Jlassi · M. E. Hamza · A. Sahli (✉)
Université de Carthage, Institut National Agronomique de Tunisie (INAT), Tunis, Tunisia
e-mail: sahli_inat_tn@yahoo.fr

W. Abdallah
e-mail: abdallah1wajdi@gmail.com

M. Allani
e-mail: allani.mohamed@gmail.com

F. Faidi · Z. Daouthi · A. Amara · H. Selmi
Water Users Associations: GDA Sidi Mansour, Sidi Saad, Touila, Fjij, Kairoaun, Tunisia

A. Zouabi · K. Selmi · T. Ayoub · R. Béji · M. A. Trabelssi · F. Joumade-Mansouri
E. Chalghaf · F. Stoffner
Commissariat Régional au Développement Agricole de Kairouan (CRDA), Kairouan, Tunisia

W. Abdallah · M. Allani · R. Mezzi · R. Jlassi · A. Romdhane · H. W. Müller
Bundesanstalt für Geowissenschaften und Rohstoffe (BGR), Hannover, Germany
e-mail: JohannesWerner.Mueller@bgr.de

© Springer Nature Switzerland AG 2019
A. Froehlich (ed.), *Embedding Space in African Society*, Southern Space Studies,
https://doi.org/10.1007/978-3-030-06040-4_4

objectives were (1) to create a specific GIS data base for the four irrigated districts of the area (Sidi Mansour, Sidi Saad, Fjij and Touila) based on the characteristics of cultivated crops, soil types and used irrigation systems; (2) to assess spatial and temporal variation of soil water budget terms from plot and farm levels to irrigated district and regional scales; (3) to map results for different time steps. The achievement of these objectives was made possible using the WEAP-MABIA Model. Thus, daily Penman–Monteith reference evapotranspiration (ETo), effective precipitation (PE), crop water requirement (CWR), actual crop evapotranspiration (ETa) and irrigation water requirement (IWR) were estimated for the four irrigated districts using spatially distributed parameters on climate, crop, soil characteristics, irrigation system and basic irrigation management practice during the cropping season 2014/2015. The delivered information is maps of the Sidi Saad Dam System with its four irrigated districts and their related farms and plots; representing the current land use, the water consumption at farm level; the crop water requirement (CWR) and the irrigation water requirement (IWR) at a daily, weekly, monthly and yearly steps. Also and thanks to WEAP (Water Evaluation And Planning system) tool functionalities, these results can be displayed on Google Earth and shared between all water irrigation managers.

Keywords

GIS · MABIA-WEAP software · Irrigation water managment · Water users associations · Crop and irrigation water requirements · Plot-Farm-WUA-Regional scales

1 Introduction

The rational management of water resources in agriculture is becoming more and more a crucial issue for the environmental and economic sustainability of the primary sector especially in the regions where water is scarce and should be saved for other uses. Efficient management of land and water resources in irrigated agriculture requires comprehensive knowledge on many variables including climate, soil, land use, crops, water availability, water distribution networks, management practices, etc. Most of these data are spatially distributed and their integration and use in irrigation planning and management has promoted the widespread utilization of Geographic Information Systems (GIS) and other modern information technologies (Todorovik and Steduto 2003; D'Urso et al. 2013).[1] In fact, GIS allows users to exchange and combine geo-referenced information coming from different sources and to integrate those data with models and decision support tools.

---

[1]Todorovik, M and Steduto, P 2003, 'A GIS for irrigation management', Physics and Chemistry of the Earth, vol. 28, pp. 163–174.

The purpose of this work is to present the development, operational functionalities and spatial modeling applications of a GIS-based irrigation water management system, to provide farmers, Water Users Associations and Local Authorities in Kairouan region with spatial and temporal information on Crop Water Requirement (CWR) and Irrigation Water Requirement (IWR). This work was implemented within the framework of the CREM-BGR project, funded by the German Cooperation and aimed to strengthen water management in Maghreb countries as well as to introduce innovative approach and tools of strategically managing irrigation that will result, not only in water savings across the region, but also in best practices that may be applicable to other regions facing similar resource concerns.

2   Description of the Study Area

The Sidi Saad Dam system is located in the center of Tunisia, downstream of the Oued Zroud Watershed which covers an area of 8650 km^2. The dam was constructed in 1983 to store water for irrigation and to prevent flood events in the region, specially, Kairouan city. By 1985, a first irrigation system had been developed which taps water from the dam to irrigate the 1225 ha of Sidi Saad public irrigated area. By 1993, an extended irrigation system was created to distribute water for three supplemental public irrigated areas, Touila (950 ha), Sidi Mansour (1718 ha) and Fjij (789 ha).

The area experiences semi-arid climatic conditions with an annual rainfall ranges from 100 to 335 mm and an annual reference evapotranspiration ETo of 1330 mm.

Agricultural and livestock productions are the main economic activities in the region. The main agricultural crops are olive-trees, barley, winter wheat, oat, alfalfa, tomato and green pepper.

At the level of each irrigated district, water management and distribution are ensured by a Water Users Association, which is responsible of the water distribution and billing farmers. In the recent past, the regional administration (CRDA) uses a certain criterion for fees collection from Water Users Association and consequently from farmers based on their irrigated area and a gross irrigation water requirement for cultivated crops. In the present context of water shortage, this criterion of flat rate per hectare was reviewed and flow meters were installed at networks header of each public irrigated area and at each operating hydrant. But these solutions from the administration have yet to get the approval of farmers. As a result, a great effort is needed to ensure efficient utilization and sustainable use of irrigation water through improved water management (Abdallah et al. 2016a).[2]

---

[2]Abdallah, W, Smaoui, Y, Zahaf, A, Sghaier, M, Romdhane, A, Brini, R, Tmimi, A, Selmi, K, Ayachi, N, Joumade-Mansousi, F, Harrabi, M, Béji, R, Trabelsi, MA, Ayoub, T, Chalghaf, E, Hamza, ME, Sahli, A, Müller, HW 2016a, Document Technique: Etablissement d'un diagnostic des ressources en eau et de leur utilisation dans le secteur agricole dans la région de Sidi Saad - Kairouan, CRDA Kairouan - Projet CREM-Volet BGR, Tunis.

3 Methodology

3.1 Development of a GIS Based Data Base

The use of GIS to design an integrated irrigation water management service requires a procedure for adequate depiction of the system. The data base represents a combination of informations coming from different sources and formats. The GIS data base has to be composed of different thematic layers and then according to the specific needs, the layers are integrated and used for the creation of thematic maps.

3.1.1 Collection of Data
The collected data include, irrigated districts limits, farms and plots limits, water network transfer, physical soil characteristics, land use and irrigation systems.

3.1.2 Scanning and Georeferencing of Existing Maps
Existing maps of water network and irrigated districts limits were scanned and imported in ArcMap. Géo-referencing was done using ground control points.

All coordinates are reported in X et Y coordinates as per the following projection parameters:

– Coordinate System: Carthage
– Projection: Universal Transverse Mercator Zone 32 N.

3.1.3 Pipe Network Database Creation
The irrigation water source is the Sidi Saad Dam and water is conveyed through two main pipes to the four Irrigated Districts of the study area. At the level of each irrigated districts water is conveyed to the farms through pipe network composed of pumping stations, different hydraulic structures and hydrants at the level of the farms.

Based on the georeferenced maps of the irrigation networks, a database of the different networks was created, contained information about pipes characteristics namely pipe length, width and type.

Concerning the hydraulics structures, their localization was done using a GPS and data concerning their characteristics were collected and implemented in the database. For the hydrants, name of user, meter ID and X & Y coordinates were added.

3.1.4 Creation of the Soil Physical Characteristics Maps
Based on 36 soil samples results of the area, an interpolation map was created using the IDW method in ArcMap. The interpolation Map allowed the determination of the spatial distribution of soils type in the irrigated districts of the study area.

3.1.5 Delineation of Farms and Plots
Based on field survey and the collaboration of the Water Users Associations of each irrigated district, the delimitation of each farm and each plot was done.

3.1.6   Plot Characteristics: Land Use and Irrigation Systems

For each delimitated plot, data regarding the land use for the growing season 2014/2015 and the irrigation system used was collected.

3.2   Decision Support System Software

The supply and demand of irrigated sector assessed at the different spatial and temporal scales require linking land cover information from land survey and GIS, with data from field monitoring and modeling. We used the Water Evaluation and Planning "WEAP" (Seiber and Purkey 2012)[3] and its MABIA module (Jabloun and Sahli 2012)[4] to calculate spatially explicit water balances, to model and map the water provisioning across the landscape, and to elucidate general patterns and changes in water demand caused by changes in climate and land cover at plot, farm, irrigation district and regional levels.

3.2.1   Water Evaluation and Planning Tool (WEAP)

The Water Evaluation and Planning (WEAP) tool was chosen to be implemented in this study. WEAP is one of the main 3.2 Decision Support System (DSS) used for Integrated Water Resource Management and planning. It offers an-easy-to use framework for water planning and scenario assessment. WEAP simulate the water demand, supply, flow, storage, etc.

The main objective of the software is to present a realistic view of the available water resource distribution. It was conceived to simulate the water supply and demand. It allow to assess water resource management policies between different sectors (Agriculture, Tourism, Industry…). The WEAP software simulates the water Balance of the demand sites and supply sources. It is based on the water balance model on a monthly basis and can be used at the level of a catchment as well as at a more complex level like Region and Country (Seiber and Purkey 2012).

Recently, WEAP has been widely used for various management problems including modeling water supply and demand at regional level, assessment of climate change impact of water resource and water irrigation management (Al Omari et al. 2015).[5]

---

[3]Seiber, J, Purkey, D (Stockholm Environment Institute, U.S. Center) 2012, SEI-WEAP-brochure, viewed 12 June 2016, http://sei-us.org/Publications_PDF/SEI-WEAP-brochure-Aug2012.pdf.
[4]Jabloun, M, Sahli, A (Bundesanstalt für Geowissenschaften und Rohstoffe (BGR) [Federal Institute for Geosciences and Natural Resources]) 2012, 'WEAP-MABIA Tutorial', 97p., viewed 12 June 2016 http://www.bgr.bund.de/EN/Themen/Wasser/Produkte/_functions/Produkte_Software_en.html.
[5]Al-Omari, AS, Al-Karablieh, EK, Al-Houri, ZM, Salman, AZ and Al-Weshah, RA. 2015, 'Irrigation Water Management in the Jordan Valley Under Water Scarcity', Fresenius Environmental Bulletin, vol. 24, no. 4, pp. 1176–1188.

3.2.2 Water Balance Model: MABIA Sub-model in WEAP

Within WEAP, different agricultural catchment calculation methods can be used. In this study, we use the MABIA crop module (Jabloun and Sahli 2012) that simulates irrigation catchment performance. The WEAP-MABIA method permits not only to look at the effects of climate change on water resources but also to assess the impacts on crop physiology. It specifically simulates daily evaporation and transpiration, irrigation requirements, crop growth and yields. The MABIA Method uses the 'dual' crop coefficient Kc method (Kc = Ke + Ks Kcb), as described in Allen et al. (1998),[6] whereby the Kc value is divided into a 'basal' crop coefficient, Kcb, and a separate component, Ke, representing evaporation from the soil surface. The basal crop coefficient represents actual ET conditions when the soil surface is dry but sufficient root zone moisture is present to support full transpiration. The WEAP-MABIA method computes daily water mass balances within each catchment considering two buckets or soil compartments. The top bucket is defined by the rooting zone and includes the surface layer (the layer that is subject to drying by evaporation). The bottom bucket is the remainder of the soil below the rooting depth down to the total soil thickness and represents the slower hydrologic response in a basin. Transpiration, evaporation, runoff, and infiltration, take place at the top bucket only, base flow is generated from the bottom bucket only. Flow from bucket one to bucket two only occurs if the bucket's field capacity is exceeded. It also allows to integrate the spatial and temporal variability of climate, soil, and crop conditions as well as water availability.

3.2.3 Conceptual Representation and Model Input

The geographical focus of the Sidi Saad Dam system WEAP model is the dam basin and includes representations of the main water management features within the watershed. This includes all of the major tributaries, the main reservoirs (Sidi Saad Dam), the groundwater aquifers (Plaine de Kairouan, El Bhira and Chrahil Nasrallah), the major irrigation pipes and the agricultural demands that are associated with this supply source.

The WEAP schematic shows, how the main features of the Sidi Saad Dam System have been disaggregated and represented in WEAP as so-called supply and demand nodes, transmission links between these nodes and water allocation rules (demand and supply priorities). Figure 1 illustrates the example of the conceptual schema, i.e. spatial layout, elaborated for Sidi Mansour irrigated district where the "farms" map was added as background. Each agricultural catchment node represents the farm in the respective part of the irrigated district. Plots and plot-specific data were implemented individually in WEAP in each node as detailed as available. Hence, the water demand analysis in WEAP will be done by the disaggregated end-use based approach of calculating crop water requirements (CWR) and irrigation water (WIR) requirements at each demand node.

---

[6]Allen, RG, Pereira, LS, Raes, D, Smith, M. 1998, Crop evapotranspiration—guidelines for computing crop water requirements, FAO Irrigation and Drainage Paper 56, Food and Agriculture Organization, Rome.

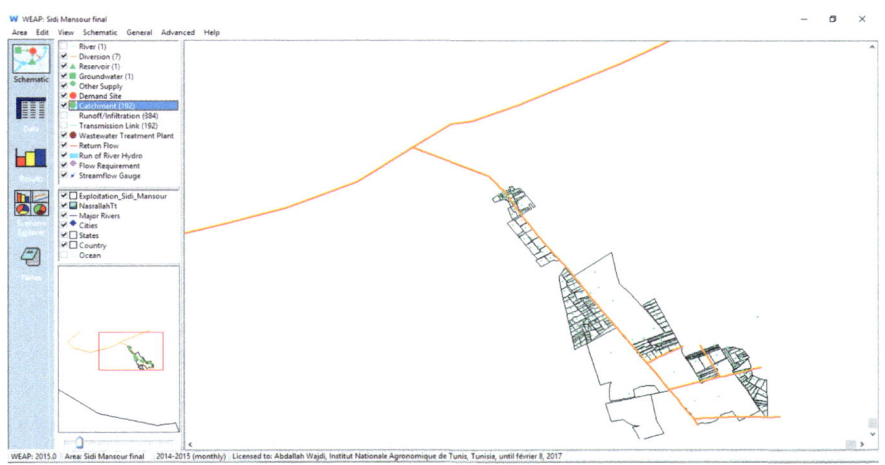

Fig. 1 Conceptual schema of Sidi Mansour irrigated district elaborated in WEAP

Simulation of the water balance by WEAP-MABIA requires input data for each plot related to the Land Use with Area and Crop (crop growth stage duration, crop coefficients, depletion factor, maximum height, rooting depth and yield response factor), the Soil (soil water capacity and soil depth), the Climate (precipitation and reference evapotranspiration) and the Irrigation (irrigation schedule and irrigation system and efficiency). Plot area and cover were based on the "Land Use" map. Daily Climate data (minimum and maximum temperature, minimum and maximum air relative humidity, sunshine duration, wind speed and rainfall) were given by the Tunisian Meteorological Agency (INM). Those data were used for the calculation of the reference evapotranspiration (ET_o) with MABIA-ETo sub-module according to the FAO-Penman–Monteith equation (Allen et al. 1998). The soil water capacity (SWAC) was obtained after processing soil particle size data from the Soil Department with MABIA-SWAC sub-module according to pedotransfer functions developed by Vereecken et al. (1989),[7] Woesten et al. (1999)[8] and Jabloun and Sahli (2006).[9] Crop–specific parameters given in MABIA-Crop Library were derived from the FAO Irrigation and Drainage Paper No. 56 (Allen et al. 1998). Special attention was paid to building crop input data, particularly planting date, crop growth stage durations, maximum height and rooting depth, as those parameters have been described to play a fundamental role

---

[7]Vereecken, H, Maes, J, Feyen, J and Darius, P 1989, 'Estimating the soil moisture retention characteristic from texture, bulk density, and carbon content', Soil Science, vol. 148, pp. 389–403.
[8]Wösten, JHM, Lilly, A, Nemes, A and Le Bas, C 1999, 'Development and use of a database of hydraulic properties of European soils', Geoderma, vol. 90, pp. 196–185.
[9]Jabloun, M and Sahli, A 2006, 'Development and comparative analysis of pedotransfer functions for predicting characteristic soil water content for Tunisian soil', 7th Tunisia-Japan Symposium on Society, Science and Technology, Sousse, Tunisia, 6–7 December 2006, pp. 170–178, TJASSST2006, Tunis.

in the model outcomes. Those data as well as irrigation systems were collected during field surveys.

Crop Water Requirement (CWR) and Irrigation Water Requirement (IWR) were estimated by considering an explicit irrigation scheduling with a timing criterion "When 90% of the Readily Available Water (RAW) in the rooting depth were consumed" and a depth criterion "Refilling Soil Water Capacity".

4 Results and Discussion

4.1 Developed GIS for Irrigation Water Management

The four irrigated districts of Sidi Saad Dam system and their limits and water distribution networks are illustrated in Fig. 2. The irrigated district Sidi Saad occupies 1671 ha distributed within 70 farmers. Farms area varies from 0.4 to 694 ha. The number of hydrants is 148. Concerning Sidi Mansour irrigated district, it occupies 2144 ha distributed within 166 farmers. Farms area varies from 0.4 to 1123 ha. The number of hydrants is 182. For Fjijj irrigated district, it occupies 826 ha distributed within 63 farmers. Farms area varies from 0.6 to 522 ha. The number of hydrants is 60. Finally, the irrigated district Touila-Sidi Kheder occupies

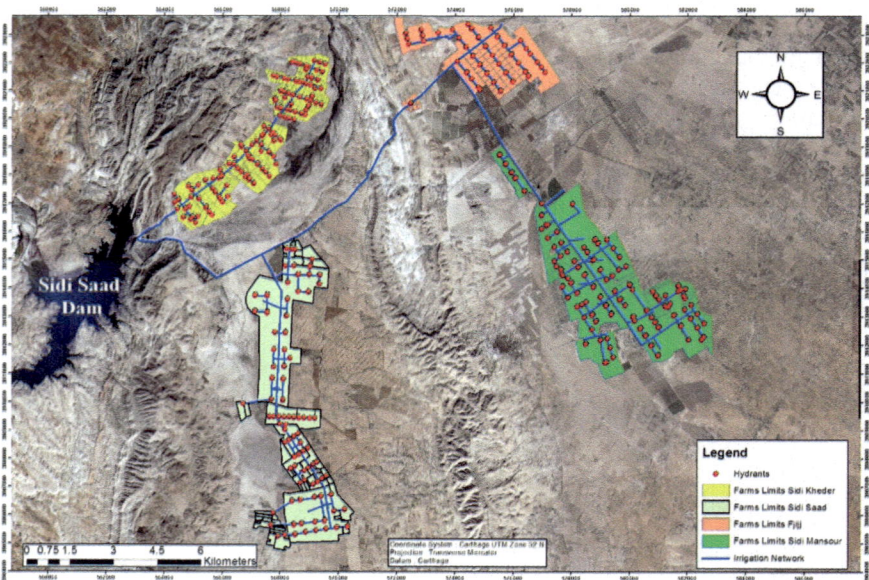

Fig. 2 Basic thematic map of the Sidi Saad Dam system irrigated districts including farms limit, water distribution networks and hydrants

997 ha distributed within 284 farmers. Farms area varies from 0.2 to 16.5 ha. The number of hydrants is 179.

In the irrigated district Touila-Sidi Kheder, the number of parcels per farm during the last growing season 2014–2015 ranged from 6 parcels and one parcel with an average 1.3 plots per farm (Fig. 3a). Analysis of the spatial distribution of crops shows that the dominated crops are olive trees with 647 ha (67% of the total area), barley with 167 ha (10% of the total area) and wheat with 37 ha, (4% of the total area). It is interesting to note that there were over 99 ha of intercropped olive groves. Fallowing is less important practice in the irrigated district with 125 ha which represent 13% of the total area.

In Sidi Saad irrigated district, the number of parcels per farm during the growing season 2014–2015 ranged from 18 parcels and one parcel with an average 2.5 plots per farm (Fig. 3b). Analysis of the spatial distribution of crops shows that the dominated crops are olive trees with an area of 452 ha (34% of the total area), barley with 479 ha (30% of the total area) and Oat with 213 ha, (13% of the total area). There were over 56 ha of intercropped olive groves. It was found that fallowing is a common and important practice in the irrigated district with 328 ha which represent 21% of the total area. For Fjijj irrigated district, the number of parcels per farm during the last growing season ranged from 11 parcels and one parcel with an average 1.2 plots per farm (Fig. 3c). Analysis of the spatial distribution of crops shows that the dominated crops are olive trees with 420 ha (52% of the total area), wheat with 35 ha (4% of the total area) and tomato with 21 ha, (2.5% of the total area). There were over 45 ha of intercropped olive groves. In this

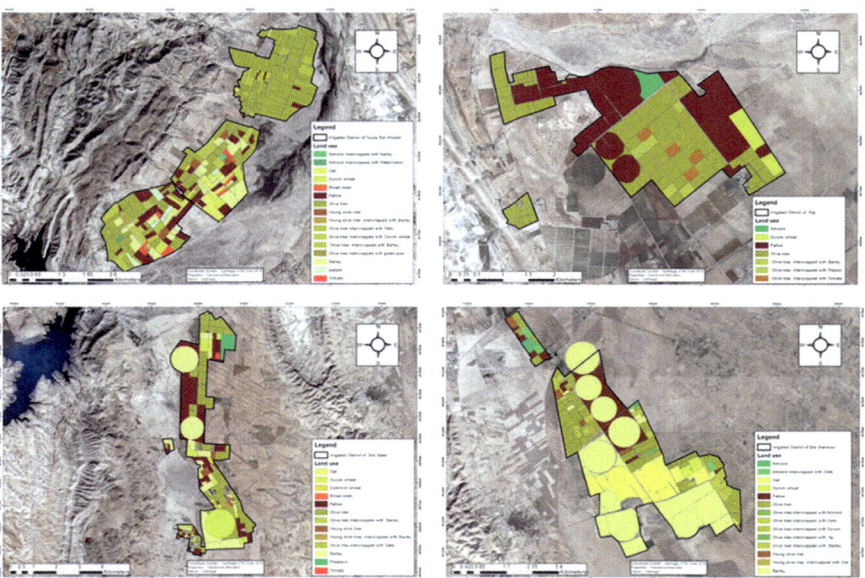

Fig. 3 Basic thematic map of the land use in the irrigated districts

irrigated district, fallowing is an important practice with 338 ha which represent 42% of the total area. Finally, in Sidi Mansour irrigated district, the number of parcels per farm during the last growing season ranged from 8 parcels and one parcel with an average 1.3 plots per farm (Fig. 3d). Analysis of the spatial distribution of crops shows that the dominated crops are barley with 908 ha (47% of the total area), olive trees with 495 ha (26% of the total area) and Oatmeal with 314 ha, (16% of the total area). There were over 99 ha of intercropped olive groves. Fallowing is less important practice in the irrigated district with 135 ha which represent 7% of the total area.

Integration of data and consequent customization of GIS environment, presented in this work, are developed for the purpose of satisfying requirements of the final users of GIS database which are, in this case, the implicated regional and local authorities as well as the Water Users Associations. The Project Team held a meeting to explain the developed GIS model and provided training to support the staff from the relevant agencies on GIS tools use. A Restitution Workshop was also organized to present and review the developed GIS to all the beneficiaries' personnel and interested stakeholders. A Technical Report of the developed GIS, all digital data and maps were prepared and distributed to all implicated administrations as part of the Validation Workshop (Abdallah et al. 2016b).[10] In addition, on request of the regional stakeholders, complementary work on the GIS model was carried out as well as ground truthing and on-the-job training to relevant administration officers in charge of GIS in Kairouan. The trained staff should be able in the future to improve/adapt/update the developed GIS.

4.2   Information on Crop Water Requirements Over the Temporal Dimension

Potential crop ET is the crop water demand, or the optimal amount of water that the crop needs to face the atmospheric water demand. It has been calculated with WEAP-MABIA, it takes into account the actual crop conditions for each day in terms of crop growth stage and climate conditions, monitored by means of daily reference evapotranspiration (ETo) data. Figure 4 shows, as an example, the profile in a parcel occupied by olive trees.

From the daily water balance, we derived, by means of the developed DSS, daily potential crop evapotranspiration maps with values expressed in mm day^{-1}. Maps obtained for the different crops were summarized to derive daily maps of crop water requirements for the whole agricultural area which can be aggregated for different time steps. Figure 5 shows, as an example, maps of Crop Water Requirements

---

[10]Abdallah, W, Smaoui, Y, Zahaf, A, Allani, M, Mezzi, R, Jlassi, R, Romdhane, A, Sghaier, M, Stoffner, F, Trabelsi, MA, Béji, R, Ayoub, T, Kouraichi, M, Chalghaf, E, Sahli, A, Müller, HW 2016b, Support Technique: Cartes thématiques de l'usage agricole de l'eau dans la zone Sidi Saad – Nasrallah. CRDA Kairouan - Projet CREM-Volet BGR, Tunis.

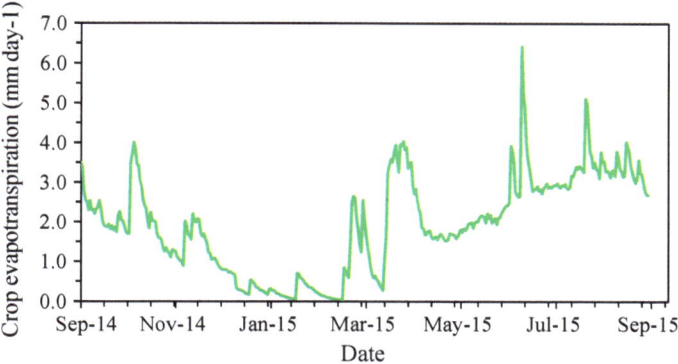

Fig. 4 Daily olive tree water requirements during the growing season 2014–2015

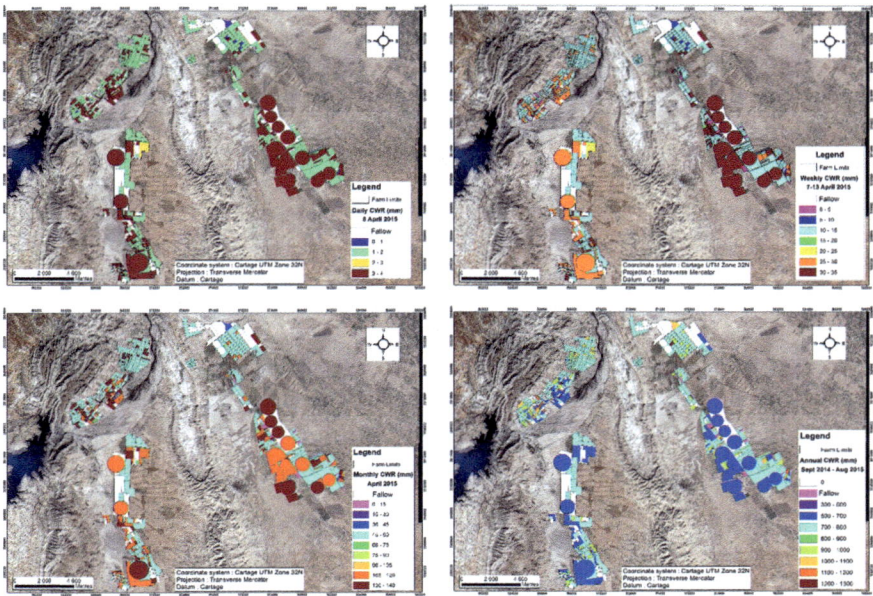

Fig. 5 Example of daily, weekly, monthly and yearly crop water requirements map

(CWR) computed on daily (8th April 2015); weekly (7th to 13th April 2015), monthly (April 2015) and Annual (cropping season September 2014–August 2015) basis.

The daily crop water needs values obtained here are consistent with those of other studies. As it was expected, parcels with intercropped almond or olive have the greatest Crop Water Requirements (from 900 in almond-water melon intercropping system to 1300 mm year^{-1} in olive-wheat intercropping system), followed by green pepper fields (1180 mm year^{-1}), almond (1100 mm year^{-1}), wheat (750 mm year^{-1}), olive (720 mm year^{-1}), pistachio (680 mm year^{-1}), tomato (770 mm year^{-1}), oat (600 mm year^{-1}) and barley (535 mm year^{-1}).

4.3 Information on Irrigation Water Requirements over the Temporal Dimension

Irrigation water requirement (IWR) is the total quantity of water applied to the land surface in supplement to the water supplied through rainfall and soil water profile to meet the water needs of crops for optimum growth, i.e. the crop water requirements (CWR). As mentioned below, it has been calculated from the daily water balance performed by WEAP-MABIA by considering an explicit irrigation scheduling with a timing criterion "When 90% of the Readily Available Water (RAW) in the rooting depth were consumed" and a depth criterion "Refilling Soil Water Capacity". It takes into account the actual crop conditions for each day in terms of soil water depletion, crop water requirements and crop sensitivity to water stress, rooting depth, soil water capacity and climate conditions (reference evapotranspiration and rainfall). Figure 6 shows, as an example, the daily soil water depletion in a parcel occupied by barley crop when the proposed irrigation scheduling criteria were applied.

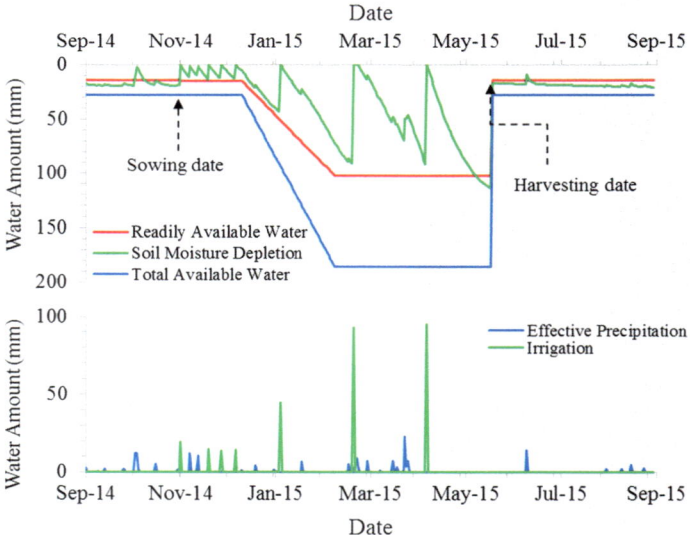

Fig. 6 Example of irrigation water requirements determination for barley crop according to the daily water balance in its rooting depth

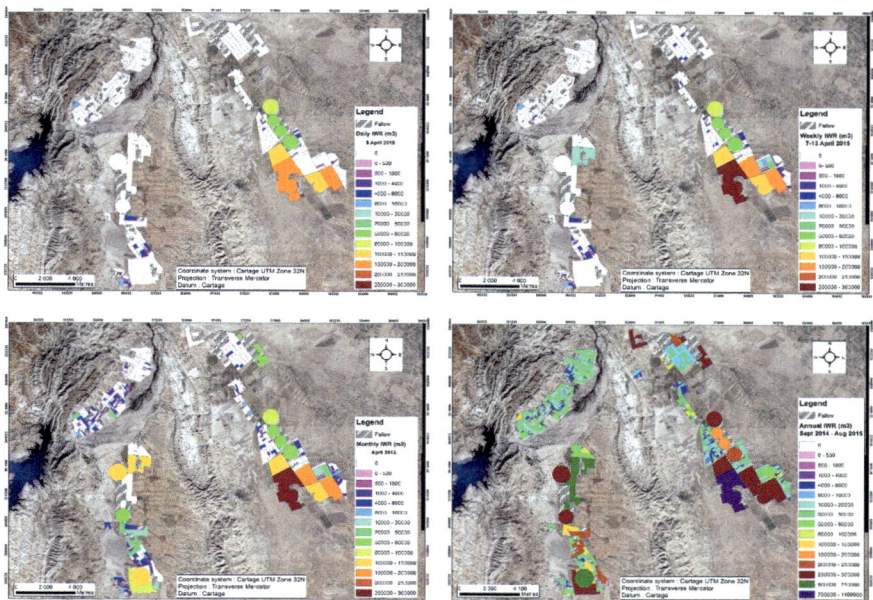

Fig. 7 Example of daily, weekly, monthly and yearly Irrigation Water Requirements map

In fact, under the climate conditions of the growing season and to avoid that barley experience water stress, seven irrigations were needed: four during the initial stage, one during the development stage and two during mid and late season stages. For each irrigation, the amount of applied water was calculated depending on root depth and soil water capacity.

From the daily water balance, we derived by means of the developed DSS daily irrigation water requirements maps with values expressed in mm day^{-1} or in m^{-3} day^{-1} by considering the parcel area (1 mm day^{-1} corresponds to 10 m^{-3} ha^{-1} day^{-1}). Maps obtained for the different crops were summarized to derive daily maps of irrigation water requirements for the whole agricultural area which can be aggregated for different time steps. Figure 7 shows, as an example, maps of Irrigation Water Requirements (IWR) at farm level computed on daily (8th April 2015); weekly (7th–13th April 2015), monthly (April 2015) and Annual (cropping season September 2014–August 2015) basis.

The daily crop irrigation water needs values obtained here are consistent with those of other studies conducted under the same pedo-climatic conditions. As it was expected, Irrigation Water Requirements (IWR) vary with species, climatic and soil conditions and with the growth period and cropping system (intercropping or monoculture). Then, farms with intercropped parcels and/or summer vegetables (green pepper, tomato water melon,) and/or fruit trees (almond, olive, pistachio) have the highest Irrigation Water Requirements in comparison to those with cereals or oat in mono-culture. For example, barley has the smallest IWR with value around 250 mm for the growth period November 2014–May 2015. On the contrary, IWR of green pepper were found to have the largest values with 1035 mm for the growth periods September 2014–November 2014 and April 2015–August 2015.

The suggested methods for daily IWR estimation during the cropping season give a very good indication of its spatial distribution and can be used directly in irrigation alert by the Water Users Association and in irrigation water management by the local and regional Authorities. The reason behind this is the wide fluctuation of CWR and IWR from day to day, depending on meteorological conditions, availability of water, land use and plant growth.

5 Conclusion

Irrigation water management is a complex task since it depends upon various factors such as land use, crop growth stage and parameters, climate and soil type data. Efficiency depends upon the reliability of these data. For this, the presented work covers two topics relevant to the improvement of irrigation management in Sidi Saad Dam System: The development of a local GIS irrigation database and the estimate of crop water requirements and irrigation water requirements, by mean of a dedicated WEAP-MABIA DSS. Combining these two tools has been embedded in an integrated methodology, which will be operational since it is implemented according to request of local water managers and the regional stakeholders. Besides, this latter accepts input from multiple data sources and offers options for updating data and including alternative methods. The delivered information is maps of the Sidi Saad Dam System with its four irrigated districts and their related farms and plots; representing the current land use, the crop water requirement (CWR) and the irrigation water requirement (IWR) at a daily, weekly and monthly steps. Also and thanks to WEAP tool functionalities, these results can be displayed on Google Earth and shared between all water irrigation managers (results not shown) to improve the irrigation scheduling, and accordingly, to make a better decision about which field (or crops) to irrigate, when and how much.

Author Biographies

Wajdi Abdallah is a Master in Sustainable Management of Water Resources at National Institute of Agronomy, Tunis. Under the guidance of Dr. Ali Sahli, he received an Engineering degree from National Institute of Agronomy, Tunis in 2014, and a graduated professional master's degree in Irrigation and Drainage from the National Institute of Agronomy, Tunis in 2016. His research focuses on irrigation management and optical and radar remote sensing.

Ali Sahli is Associate Professor and Director of Studies of National Institute of Agronomy, Tunis, where he has been since 1995. From 2003 to 2017 he served as Assistant Professor of agricultural higher education. During 1995–2003 he served as Assistant of agricultural higher education. He received an Engineering degree from National Institute of Agronomy, Tunis in 1988, a graduated engineering degree from the National Polytechnic Institute of Toulouse and a Diploma of Advanced Studies from the National Polytechnic Institute of Lorraine. He received his PhD in mechanical and energetic from the National Polytechnic Institute of Lorraine in 2001. From 1991 to 1993 he worked at the University Institute of Technology at the University of Nancy 1 as a temporary substitute teacher, and from 1993 to 1995 he worked at the European School of Engineering in Materials Engineering, National Polytechnic Institute of Lorraine as a temporary substitute teacher. His research interests heat and mass transfers with application on environmental physics, bioclimatology and remote sensing.

Water Quality Information for Africa from Global Satellite Based Measurements: The Concept Behind the UNESCO World Water Quality Portal

Thomas Heege, Karin Schenk and Marie-Luise Wilhelm

Abstract

Freshwater as one of the most relevant resources for life is facing increasing human made pressures. Suitable information about the status of the water quality in lakes and rivers is sparse, although required for environmental assessments and impact monitoring: There is a vast demand on actual data in many countries, where water policies and management decisions are based on scarce and unreliable information. Satellite data with newest data analytics technologies can already contribute to this today with regular mapping and monitoring in freshwater systems: Consistent information of valuable water quality products are derived for single applications in small lakes, covering extended river basins or the whole world, as provided by the UNESCO World Water Quality Portal. A number of examples from this first global water quality portal is discussed, addressing ecological and economic issues in Africa. At the conceptual level, UNESCO and EOMAP advocate the long-term consistency of the data of these new measurement capabilities: Both satellite sensing and data processing technologies are rapidly evolving. Hence, nowadays concepts should already ensure that the data products are globally intercomparable and in future, even if the accuracy of information products become better and better. This ensures that the sustainable development goals can be supported with meaningful, comparable indicators over time.

T. Heege (✉) · K. Schenk · M.-L. Wilhelm
EOMAP GmbH & Co. KG, Seefeld, Germany
e-mail: heege@eomap.de

K. Schenk
e-mail: schenk@eomap.de

M.-L. Wilhelm
e-mail: wilhelm@eomap.de

© Springer Nature Switzerland AG 2019
A. Froehlich (ed.), *Embedding Space in African Society*, Southern Space Studies,
https://doi.org/10.1007/978-3-030-06040-4_5

81

1 Introduction

Freshwater resources are limited and face rising pressures due to increased population, exploitation of environmental resources, or effects through climate changes. The OECD[1] (2012) states that, for example, the deterioration in water quality resulting from eutrophication has reduced biodiversity in rivers, lakes and wetlands by about one-third globally. With the largest losses observed in China, Europe, Japan, South Asia and Southern Africa, there is an urgent demand to enhance status information on water quality at local and global level. Despite having experienced more than 10 years of continuous economic growth, Africa today faces great water resource management challenges. With 10% of the world's renewable water resources, more than 60 trans-boundary basins, a low level of water development and utilization and increasing population, Africa's future economic growth will continue to be constrained by the development of its water resources. Today, in many African countries, water policies and management decisions are based on sparse and unreliable information.[2]

In this challenging context, it is not surprising that the actual agenda of governments demonstrate a paradigm change and a new era: Agencies and industry require improved environmental data, develop their directive monitoring obligations, set up socio-economic full-cost calculations and need to understand upcoming risks through new areas of conflicts over clean water and food.

The targets of the Sustainable Development Goals (SDG) address these and further efforts on sustainability of human life. The anticipated indicators to monitor the progress towards the SDGs with statistical data are under continuous enhancement to include globally tangible measures using new technologies. UNESCO, UNEP[3] and environmental agencies believe the answers should also entail the resources of satellite-based Earth Observation measurements (EO): A comprehensive range of satellite-based water quality parameters in surface waters can be measured: Turbidity, Chlorophyll as indicator for the trophic status, and the appearance of harmful Cyanobacteria bloom indicators.

These measurements from space provide unique holistic environmental information over our connected water systems which are not available by any other water analytics method: Area-wide, going back more than 30 years, and nowadays with a spatio-temporal resolution which corresponds better to natural dynamics than any other sampling method. However, the new method comes with restrictions: It can only provide measures from the first meters penetrated by light, the trophic zone, and only under cloud-free conditions. The accessible water quality parameters

---

[1]OECD, 2012. Water quality and agriculture – meeting the policy challenge. OECD Studies on Water. OECD Publishing.
[2]Earth Observation for Water Resource Management in Africa
  Benjamin Koetz, ZoltánVekerdy, Massimo Menenti and Diego Fernández-Prieto (Eds.), January 2016.
[3]UN-Water 2016, Towards a Worldwide Assessment of Freshwater Quality: A UN-Water Analytical Brief 2016, Published by UN-Water, November 2016. Available at http://www. unwater.org/publications/towards-worldwide-assessment-freshwater-quality/.

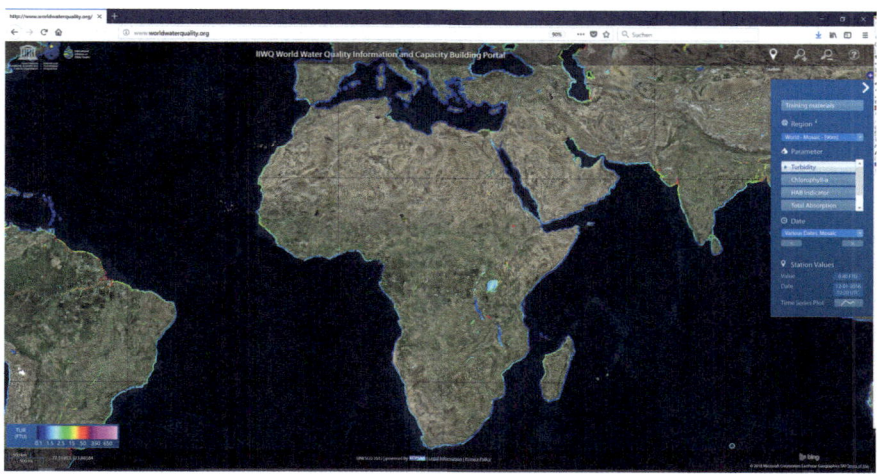

Fig. 1 The UNESCO IIWQ World Water Quality Portal (www.worldwaterquality.org). At the right side of the portal a number of tools are provided in the blue bar. Users can select parameters for visualization, use tools to access values for self-defined virtual stations or access training documents

need to be related to optically active water constituents, with spectrally distinct absorption or scattering features in the visible range of the light. As for the wide range of optical conditions, waterbodies require complex physics based algorithms to convert the spectral measurements of the satellite instruments into globally consistent and comparable measures. Using operational data analytics and an unmatched cost efficiency, the new method provides measurements from over thousands of river kilometres, millions of virtual sampling stations and lakes combined (Fig. 1).

2 Understanding the Value of Satellite Observations: The UNESCO IIWQ Portal

The UNESCO International Initiative on Water Quality (IIWQ)[4] addresses capacity building for satellite-based water quality monitoring technologies, especially for remote areas and in developing countries such as Africa where water quality monitoring networks and laboratory capacity lack. Developed by EOMAP, a world leading specialist for aquatic earth observation monitoring services, UNESCO launched its World Water Quality Portal (www.worldwaterquality.org) in 2018.

---

[4]The International Initiative on Water Quality (IIWQ) of UNESCO's International Hydrological Programme (IHP) Online, Available at https://en.unesco.org/waterquality-IIWQ.

The comprehensive portal assists governmental institutions and industry with global water quality assessment for streams, lakes, and rivers providing direct and free access to up-to-date space-based measurements for any location in inland and near-coastal surface waters.

The portal allows users to instantly obtain measurements at freely selectable virtual stations for any location worldwide. The underlying concentrations are provided straight instantly in the value box for the selected parameter, place and time. It provides a range of satellite-based water quality parameters such as turbidity, chlorophyll, and indicators for toxic Cyanobacteria blooms. It also includes functionalities to select different time periods: Historic measurements are provided at a 30 m resolution for selected regions of each continent throughout 2016, and can be continued with various spatial and temporal resolutions for every country. For example, reports on a yearly tropic status based on chlorophyll can be directly exported for any user defined station within selected time series regions—with only one mouse click away. Figure 2 provides an example of the spatial distribution of Chlorophyll for such a virtual station in the central Egyptian part of Lake Nasser. The spatial distribution at this date in March 2016 already varies by a factor of 5 in this part of the reservoir with impressive dynamics for the different fractal shaped bights. The annual readings as well show high temporal dynamic and seasonal changes of a factor of 10. These readings are provided after selecting the time series plot in Fig. 3. The easy-to-access report classifies this part of the reservoir as predominant mesotrophic at around 5 µg/l for the mean Chlorophyll concentrations.

The online interface, based on satellite-derived information, gives users access to an easy-to-use tool providing detailed global water quality information and helps to understand the diversity of space-based measurements in space and time. But the

Fig. 2 Spatial distribution of phytoplankton and its key pigment Chlorophyll in central Lake Nasser/Egypt in March 7, 2016. The white pointer shows the location of the selected virtual station. The actual water quality value and unit is provided in the lower section of the blue bar

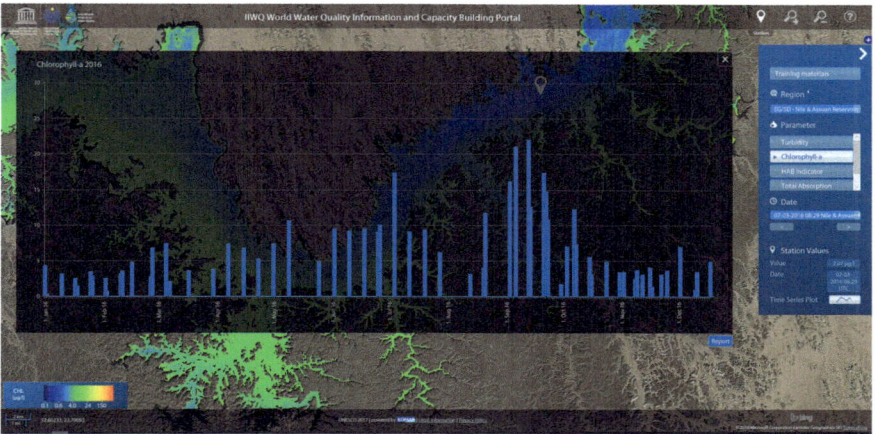

Fig. 3 Temporal dynamics of Chlorophyll at a selected virtual station in central Lake Nasser/Egypt in 2016

World Water Quality Portal will also be a key element in improving awareness, capacity building and acceptance of EO products on a global scale.

3 The Technical Development: Satellite Sensors and Space-Based Observations

Satellites with remote sensing instruments have been monitoring the earth since 1985, when the first US Landsat satellites[5] were launched. Since then, these data are setting the basis for vast investigations of the changing environments worldwide. Between 2010 and 2020 the monitoring capabilities improved significantly in all directions – sensitivity, spectral, spatial and temporal resolution. Raised public and private investments resulted in dense and regular global earth observations. Nowadays, large capabilities are provided through the EU Sentinel satellites[6] and numerous commercial satellite systems: All satellites are equipped with different technical specifications and objectives, but many of these are capable to provide water analysis. African countries also invest significantly into earth observation, e.g. South Africa orbited its first satellite in 1999, Nigeria has launched several multimillion-dollar satellites since 2003, and Morocco launched Africa's first high-resolution imaging satellite in 2017.[7]

---

[5]Landsat Missions, United States Geological Survey (USGS) & The National Aeronautics and Space Administration (NASA), Available at https://landsat.usgs.gov/.
[6]Sentinel Online, European Space Agency (ESA), 2000–2018, Available at https://sentinel.esa.int/.
[7]Firsing, Scott (2015), Africa and space: the continent starts to look skyward, The Conversation UK Online, Available at https://theconversation.com/africa-and-space-the-continent-starts-to-look-skyward-41336.

In the past years, the data analytics and processing technologies evolved to generate usable information on water quality in lakes and river using optical multispectral sensors. Data processing technologies are of equivalent importance as sensors, as the characteristics and quality of the derived information products are defined by a package of used satellite sensors and data analytics software. Today, those consolidated packages of satellite sensors, data analytics and online service provision systems form a new type of further innovating environmental services.

The UNESCO portal is built on such a system package: Data from different satellite sensors, Landsat-8 and Sentinel-2, are transformed into global water quality measurements and continuously available through the portal stored in a geospatial data base and accessible online via the user-friendly web application.

4   The Advances of Satellite Data Analytics: Globally Harmonized Water Quality Measurements

Multispectral satellite sensors are capable of measuring water constituents using sunlight as it penetrates the atmosphere and waterbody. This light is absorbed and scattered as a function of the particles and dissolved materials in the waterbody. The reflected light spectrum detected by the satellite sensors can be used to analyse the optically active water constituents. In other words: The water colour is used to derived water quality information, as indicated in Fig. 4. However, the satellite signal is strongly modified by a number of further very variable impacts. These originate from varying atmospheric aerosols, water surface reflections, scattered light from adjacent land areas, and the observation geometry. The most accurate correction of all these impacts is thus consequentely a fundamental requirement of the satellite data analysis software.

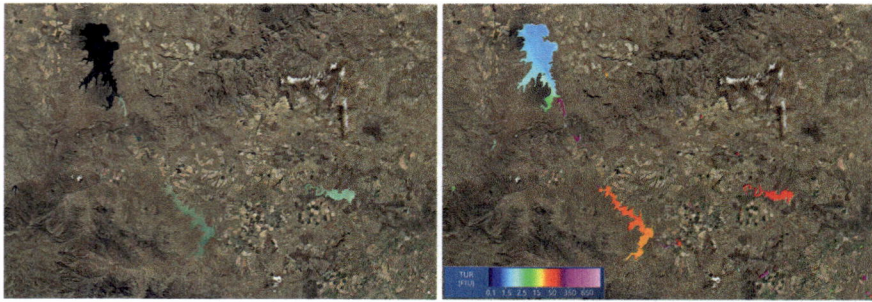

Fig. 4 The water colour is linked to the water constituents through their specific spectral absorption and scattering features. The left hand side shows the water colour as visible in a Sentinel-2 satellite record, and the right hand side shows the related turbidity for the different levels of particle scattering in the various water bodies in South Africa. The upper left is the Sterkfontein dam with low turbidity levels between 0, 9 and 5 NTU

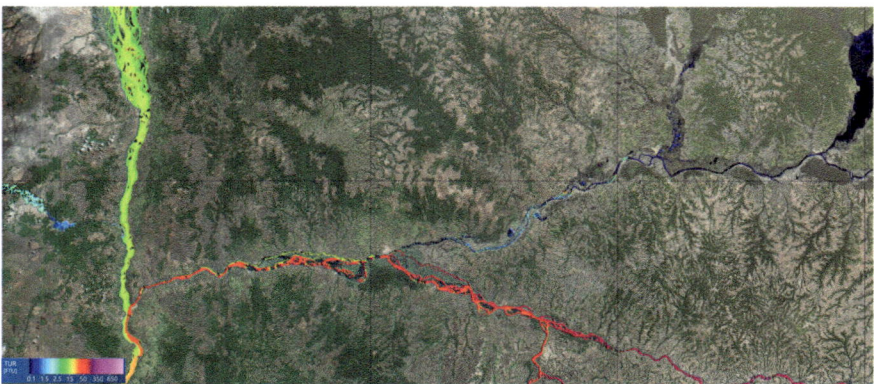

Fig. 5 Section of the Congo river basin, showing the turbidity through suspended sediments in the Congo, the Kasai river (right hand-side) and Levini river (left hand) entering the Congo

Furthermore, the main strengths of satellite-based measurements can only be exploited under the conditions demanded by the UN-water analytical brief 2016[8]: Provide consistent and reproducible products based on high-scientific standards to ensure product quality and independency on ground-truth data. These requirements can be fulfilled with physics-based data processing methods: Derived water quality information measures and their physical units relate directly to the absorption and scattering properties of the water body. They are physical properties, therefore globally comparable and independent of specific algorithms or sensors. For example, turbidity is linearly related to the physical process of the backscattering of light, while Chlorophyll and organic components relate likewise to its absorption.

UNESCO applied such a physics based data analysis technology, based on the Modular Inversion and Processing System MIP. The systematic development started more than 20 years ago at the German Aerospace Center, and has been further developed by EOMAP since 2006.[9] Today, it forms the scientific and methodological base for the variety of routine water monitoring programs from space, ranging from multi-year regulatory monitoring of the trophic status, dredging impact monitoring for the offshore industry, or environmental impact analysis after natural disasters such as the Rio Docedam collapse in Brazil. The common requirement for the very different use cases is again the consistency of the water quality information products over space and time and the different utilized satellite sensors. Bearing this in mind, historical reviews covering several satellite

---

[8]UN-Water 2016, Towards a Worldwide Assessment of Freshwater Quality: A UN-Water Analytical Brief 2016, Published by UN-Water, November 2016. Available at http://www.unwater.org/publications/towards-worldwide-assessment-freshwater-quality/.

[9]Dörnhöfer, K., Klinger, P., Heege, T., Oppelt, N. (2017): Multi-sensor satellite and in situ monitoring of phytoplankton development in a eutrophic-mesotrophic lake. Science of the Total Environment 612C (2018) pp. 1200–1214 DOI information: https://doi.org/10.1016/j.scitotenv.2017.08.219.

generations are possible as well as highest temporal resolutions combining different current sensors. Taking this fact into account, historical reviews over several generations of satellites as well as highest temporal resolutions combining different current sensors are possible.

5 The Benefits of Space-Based Measurements: Selected Use Cases

The huge African continent incorporates an endless number of freshwater issues but also great advancements within thousands of interconnected catchment areas. Space-based measures can support to monitor and evaluate these. It can provide data that wouldn't be available otherwise, support political decisions on information rather than on guesses or unaffordable surveys.

Highlighted below are some interesting aspects visualized with data that is already accessible through the UNESCO IIWQ World Water Quality Portal.

5.1 Assessing the Economic and Ecological Impacts of Human Interventions on the Sediment Balance

Figure 5 shows a section of the Congo river basin, with the Congo river flowing from the north to the south direction. The Kasai river is entering the Congo from the right hand side with high amounts of suspended matter, causing increased turbidity in the downstream Congo river. The mixing processes of the two water bodies can be followed for 100 km downstream, with the less turbid water on the western side of the river. On the western side of the figure, the Levini river is entering into the Congo, with much lower amounts of sediments: It is obvious that the Imboulou dam traps the sediments from the Levini river. The same effects can be observed at numerous dams worldwide. The spatial-temporal dynamic of turbidity and suspended sediments is accessible with time series data sets in the portal, e.g. for the Aswan Reservoir.

Figure 6 shows the impact of the basin trapping the sediment plume of river Nile in the upstream part between Sudan and Egypt. The concentration drops from almost 500 NTU to less than 5 NTU in this part of the reservoir in August 2016. This dramatic drop implies that the suspended particles are sedimented and removed from the sediment balance of the river. Such satellite-based measurements can therefore be used to quantify the sediment loss through dams[10] or for the economic and ecological evaluation of new hydropower developments over extended river systems.[11] In general terms, the sediment balance, sediment trapping

---

[10]Heege, T., Klinger, P. 2016: Space based monitoring of the sediment balance and water quality in reservoirs. Proc. Hydro 2016 conference, Montreaux 10.-12.10.2016, p 1–7.
[11]Heege, T., Kelleher, D. 2018: Reducing economic risks in hydropower developments through independent satellite based turbidity and sediment measurements in the river systems of Georgia. Proc. Hydro 2018 conference, Gdansk 15.–17.10.2018, pp15.

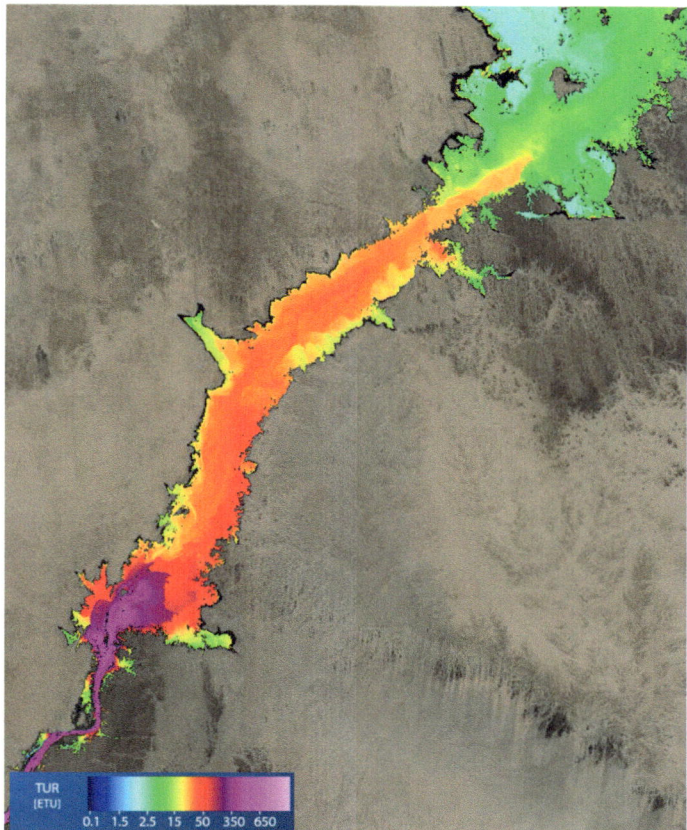

Fig. 6 River Nile inflow into the Aswan Dam on 20 August 2016, with large turbidity gradients from south to north

as well as water quality and their seasonal or long-term changes are of highest economic and ecological relevance for the lifetime of reservoirs,[12] the operation costs, but also for their river basins and connected deltas.[13] Sustainable long-term solutions to reduce those significant risks related with the sediment balance in the river basin including the reservoir and deltas are required,[14] and can now be efficiently supported through satellite-based monitoring.

---

[12]UNESCO 2011, "Sediment Issues & Sediment Management in Large River Basins Interim Case Study Synthesis Report", IRTCES 2011 Available at: www.irtces.org/isi/isi_document/2011/ISI_Synthesis_Report2011.pdf.

[13]Giosan L., Syvitski J., Constantinescu S. & Day J. 2014, "Protect the world's deltas". Nature 516, 31–33 (2014) https://doi.org/10.1038/516031a.

[14]Tessler Z.D., Vörösmarty C. J., Grossberg M., GladkovaI, Aizenman H., SyvitskiJ. P. M., Foufoula-Georgiou E. 2015, "Profiling risk and sustainability in coastal deltas of the world". Science 2015, Vol 349 (6248), 638–643.

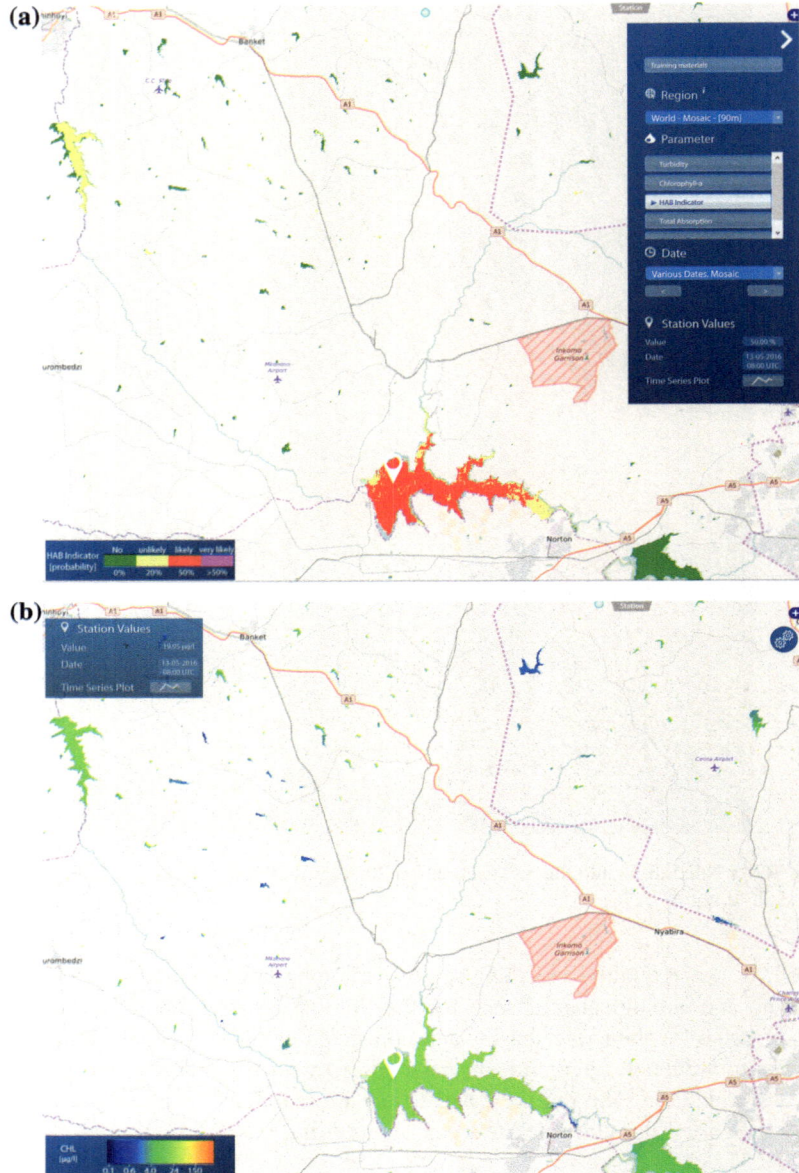

Fig. 7 Harmful algae bloom indication **a** and Chlorophyll concentrations **b** for Lake Manyame and the lakes surrounding the Manyame basin in Zimbabwe, May 13 2016

5.2 Environmental Assessment of Lakes: Trophic Status and Cyanobacteria Blooms

Satellite monitoring services are increasingly used by water agencies to evaluate the trophic status of lakes, using Chlorophyll measurements as shown in Figs. 2 and 3 for Lake Nasser, or as demonstrated with the report functionality provided by the UNESCO portal. In many regions, only a fraction of the relevant water bodies can be accessed with traditional in situ data, with few point measurements per year. In contrast, the satellite-based technology can provide typically between 10 and 100 records per year area-wide. It provides insights of the spatio-temporal dynamics and detects impact sources which could not be traceable with other methods. For agencies with obligations to assess the status of hundreds to thousands of surface water bodies, it offers numerous opportunities to improve their assessments: Cost-efficient compliance with regulatory obligations, support for policy makers for better decision making, SDG reporting, but also earlier warnings on changing environments. For example, Cyanobacteria are of increasing concern in freshwater systems worldwide, favoured by anthropogenic eutrophication and rising temperatures.[15] When they appear in masses on surface waters, they can cause immense damages on fish life, drinking water and with harmful impact human life. Lake Manyame near Harare is an example for the increasingly polluted inland water bodies in Zimbabwe, and known for the appearance of the Microcystis Cyanobacteria.[16] The well investigated lake provides also a good test bed for remote sensing researchers in Africa, e.g. Muchini et al. 2018.[17] The UNESCO portal visualizes the situation of an extended Cyanobacteria bloom in Lake Manyame clearly with the Harmful Algae Bloom indicator HAB in Fig. 7. The lake is surrounded by numerous under-investigated smaller lakes, and those well-resourced missing information can be provided by the portal for adequate decision-making about the landscape.

6 Outlook

Satellite-based monitoring of freshwaters became a global applicable technology, suitable to map global challenges and support local solutions. The technology can contribute to provide the required environmental information that is needed to address economic and ecological relevant water issues in Africa.

---

[15]Paerl HW, Paul VJ (April 2012). "Climate change: links to global expansion of harmful cyanobacteria". Water Research. 46 (5): 1349–63. https://doi.org/10.1016/j.watres.2011.08.002. PMID 21893330.
[16]Sibanda Tendaupenyu, Pamela. (2012). Nutrient limitation of phytoplankton in five impoundments on the Manyame River, Zimbabwe. Water SA. 38. 97–104. https://doi.org/10.4314/wsa.v38i1.12.
[17]Muchini, R., Gumindoga, W., Togarepi, S., Masarira, T. P., and Dube, T.: Near real time water quality monitoring of Chivero and Manyame lakes of Zimbabwe, Proc. IAHS, 378, 85–92, https://doi.org/10.5194/piahs-378-85-2018, 2018.

Despite the global applicability, it must be noted that the individual demand on service specifications differs significantly over the various user organizations, governments and water managers. Aspects such as temporal or spatial resolution, product accuracy, data aggregation, reporting details are crucial to fulfil actual demands. Still, the growing demand is the main driver for further innovations and improvements of market driven environmental services. Better algorithms and sensors provide more accurate data, more detailed parameters and higher spatio-temporal resolutions year after year.

Users can benefit the most from these rapid developments in consideration of the required standards: Globally intercomparable measures for satellite-based water quality parameters are relatively easy to define when linked to their physical inherent optical properties. Such measures are independent of the specifications of previous or upcoming satellites, and of specific data analytics technologies. Hence, users will benefit from long-term consistent data. Such intercomparable environmental data, generated by many countries and contributors, can be shared through intergovernmental platforms such as GEOSS, accessed through various front ends such as the UNESCO IIWQ portal. Finally, they provide suitable data to monitor the success of the Sustainable Development Goals with long-term comparable measures and indicators.

Author Biographies

Dr. Thomas Heege, CEO and founder of EOMAP, has more than 25 years experience in aquatic remote sensing and technical consultancy. As a remote sensing expert with experience in projects around the globe he is familiar with implementing users needs into practicable solutions for water industries and governmental institutions. Prior to founding EOMAP in 2006, he worked as a scientist and project manager at the German Aerospace Centre DLR and at Geographical and Limnological Institutes in Munich and Constance. He graduated as Physicist in 1993 at the University of Konstanz, and received the PhD in Space Sciences in 2000 at the Free University of Berlin for his research on airborne remote sensing of Lake Constance.

Karin Schenk, Head of the Water Quality Department at EOMAP, has more than 10 years experience in the fields of earth observation services and applied solutions. She successfully developed commercial water monitoring services for various environmental agencies and industry clients, and led or participated in EU funded projects such as FRESHMON or SPACE-O to innovate satellite-based services. Prior to joining EOMAP in 2011, she worked for a large remote sensing company in Munich, Germany on several land focused projects and conducted statistical data analysis for BMW Germany.

Marie-Luise Wilhelm, Head of Marketing & Communications at EOMAP, has more than 20 years experience in developing and implementing strategic, comprehensive and integrated communications and public relations activities across numerous industries and research institutions, including the oil & gas, offshore, wind and marine energy industry. At EOMAP, she has responsibility for the company's brand strategy and communication activities. Prior to joining EOMAP in 2017, Marie-Luise has held various senior marketing & communications positions in the UK and Germany. She holds of Master of Business Administration from Munich University of Applied Science and is a specialized journalist covering various science-based topics.

Space-Based Financial Services and Their Potential for Supporting Displaced Persons

David Lindgren

Abstract

This paper outlines a concept for a space-based financial services provider that could be developed in support of the 2030 Agenda for Sustainable Development in Sub-Saharan Africa. This region hosts large populations of displaced persons, which require significant resources and support from humanitarian and development organisations. In addition, many conflict-affected areas in Sub-Saharan Africa, including those that generate displaced persons, feature high numbers of unbanked people. Satellites have increased in number and capabilities due to notable advances in computing, manufacturing, and launch technologies. As such, spaced-based services, including financial services, are proving more affordable and accessible to more people than ever before in history. Coupled with the rising acceptance of cryptocurrencies, spaced-based financial services can be delivered to displaced persons to aid them in their transition from displacement to resettlement, allowing them to carry their assets with them pre, mid, and post-displacement. Humanitarian and development organisations, alongside bilateral and multilateral donors and traditional financial institutions, could prove as the initial supporters of such an initiative oriented toward achieving development results for such vulnerable populations. However, significant legal and compliance challenges, including adherence to anti-money laundering and know your customer banking best practices, pose important questions for the feasibility of a space-based financial services provider.

A Submission to the Publication on "Space embedment in African Society in regard to UN Goals 2030".

D. Lindgren (✉)
SpaceLab, University of Cape Town, Cape Town, South Africa
e-mail: davidlindgrenj@outlook.com

1   Introduction

During the United Nations (UN) Sustainable Development Summit held from 25 to 27 September 2015, UN member states adopted the 2030 Agenda for Sustainable Development, a far-reaching and wide-ranging international development initiative that contains 17 high-level targets, commonly known as the Sustainable Development Goals (SDGs).[1] As part of the 17 SDGs, they cumulatively represent 169 targets[2] and 232 associated indicators[3] that the world will use to measure progress for accomplishing the goals by the year 2030. These goals retain relevance to all UN member states in ensuring they are achieved; however, some states, particularly those in the Global South and more specifically in Sub-Saharan Africa, must achieve greater results in a relatively short duration of time as compared to their Global North counterparts, particularly in areas such as literacy, health, and economic growth and development, whereas Northern countries already possess significant advantages.[4]

1.1   Sub-saharan Africa and the Sustainable Development Goals

Sustainable Development Goal 8 and Sustainable Development Goal 10 (SDG 8 and SDG 10, respectively) remain particularly relevant to Sub-Saharan Africa and its varied, but unique circumstances relating to its colonial legacy, ongoing armed conflicts and political crises, large displaced populations, and continued underdevelopment despite significant mineral wealth and economic potential. SDG 8, which seeks to "promote sustained, inclusive and sustainable economic growth, full and productive employment and decent work for all,"[5] and SDG 10, which seeks to "reduce inequality within and among countries,"[6] directly relate to the need for increased and inclusive economic growth that enables lower levels of inequality not only within and between Sub-Saharan Africans states themselves, but also between Sub-Saharan Africa and the Global North. However, it remains important to not consider economic growth and inequality reduction in vacuums alone in and of themselves, but rather as affected and directly linked to other socio-political circumstances present in the region as well.

---

[1]United Nations, 'The Sustainable Development Agenda' (2016) Accessed July 2018: www.un.org/sustainabledevelopment/development-agenda/.
[2]Ibid.
[3]United Nations, 'SDG Indicators' (2017) Accessed July 2018: https://unstats.un.org/sdgs/indicators/indicators-list Accessed January 2018.
[4]Willige, Andrea, 'Which countries are achieving the UN Sustainable Development Goals fastest?' (2017) *World Economic Forum*, Accessed July 2018: https://www.weforum.org/agenda/2017/03/countries-achieving-un-sustainable-development-goals-fastest/.
[5]United Nations, 'Sustainable Development Goal 8' (2017) Accessed July 2018: https://sustainabledevelopment.un.org/sdg8.
[6]United Nations, 'Sustainable Development Goal 10' (2017) Accessed July 2018: https://sustainabledevelopment.un.org/sdg10.

Indeed, countries have made progress against these goals, such as for example when the UN noted in a 2017 review of SDG 8 that "From 2011 to 2014, 700 million adults became new account holders and the share of adults with an account at a financial institution increased from 51% to 61%."[7] Or, for example, the UN noting change against SDG 10 when it wrote, "From 2008 to 2013, the per capita income or consumption of the poorest 40% of the population improved more rapidly than the national average in 49 of 83 countries (accounting for three quarters of the world's population) with data."[8] As demonstrated here, progress had been achieved, and likely has since been improved upon since the periods of measurement for these results, but the global development community still has more to accomplish.

1.2 Displaced Persons and Development

Delivery of effective development results will depend on a conceptual shift and different approach to beneficiary populations if the world intends to achieve the SDGs by 2030, in particular SDGs 8 and 10. International development organisations often view the most at-risk and vulnerable populations in the world, including those fleeing conflicts, as 'objects' of development and humanitarian assistance to be served with basic goods and services, which are necessary, rather than as 'subjects' equipped with their own agency and interests beyond the immediacy of their situation. Indeed, as a report by the Norwegian Agency for Development Cooperation notes, the focus for humanitarian and development organisations "has primarily been on recipient governments rather than the local populations in villages, towns and cities that are the ultimate target group and end users of most development aid."[9] The report goes on to suggest that better development results could be achieved by development actors through increased participation of beneficiary populations by doing a "detailed and systematic analysis of the context, and a better understanding of who participates, in what activity, and for what motives."[10]

Whereas the above describes the role and importance of active beneficiaries in programme design and implementation within the international development sector, the same can be applied to the humanitarian assistance sector as well. In recognition of the failings of the humanitarian effort for supporting displaced peoples, the UN commissioned an evaluation in 2005 of its support to displaced persons, which resulted in findings that "responses to complex emergencies and disasters often

---

[7]United Nations, 'Sustainable Development Goal 8' (2017) Accessed July 2018: https://sustainabledevelopment.un.org/sdg8.

[8]United Nations, 'Sustainable Development Goal 10' (2017) Accessed July 2018: https://sustainabledevelopment.un.org/sdg10.

[9]Norwegian Agency for Development Cooperation, 'A Framework for Analysing Participation in Development' (2013) Accessed July 2018: https://www.oecd.org/derec/norway/NORWAY_A_FrameworkforAnalysingParticipationDevelopment.pdf.

[10]Ibid.

failed to meet the needs of IDPs [internally displaced persons] and other affected populations in a timely and consistent manner."[11] This same evaluation report led to reforms in the UN's approach to humanitarian support and coordination with other institutions, such as the World Bank, on assistance to displaced populations. However, despite reforms to improve, a recent World Bank report recognised that additional interventions could further strengthen the response and support provided to displaced persons by combining medium to long-term development approaches with short-term humanitarian interventions:

> The report examines available data to better understand the scope of the challenge, and reveals that the same 10 conflicts have accounted for the majority of the forcibly displaced every year since 1991, consistently hosted by about 15 countries – overwhelmingly in the developing world. The development approach focuses on tackling the medium-term social and economic dimensions of forced displacement, which complements emergency response and the rights-based agenda of humanitarian partners.[12]

As such, the current and predominant approach to displaced persons limits the long-term solutions to ensuring these populations' prosperity beyond the immediacy of the conflict, exacerbated by the short-term operating periods of humanitarian efforts versus the medium to long-term operating periods of more sustained development interventions.[13] Those affected by conflict, such as internally displaced persons (IDPs) or refugees, often leave behind lives they built where they had been accumulating some form of capital to sustain their livelihoods. As such, when they flee conflict zones, their lives' work and capital often disappears too, leaving many to start anew either in the location to which they fled or upon returning home. In either case, their capital does not follow them easily during this transient period.

Many of these populations were unbanked originally, leaving many without an option to secure their capital in the first place when they decided to flee the conflict. The World Bank report mentioned above supports this argument when it suggests to "...address long-term development challenges, as well as scale up service delivery and facilitate a transition from relying on humanitarian aid to utilizing country systems," and to "...promote self-reliance and access to jobs, [and] create opportunities through private sector investment."[14] Moreover, a recent Brookings Institution report supports the need for radical approaches to bridge the humanitarian to development approach for supporting displaced persons:

---

[11]Crisp, J., Kiragu, E., Tennant, V., 'UNHCR, IDPs and humanitarian reform' (2007) *Forced Migration Review*, Edition 29, Accessed June 2018: http://www.fmreview.org/humanitarianre form/crisp-kiragu-tennant.html.

[12]World Bank, 'Supporting Refugees and Internally Displaced Through Development' (2017) Accessed June 2018: http://www.worldbank.org/en/topic/fragilityconflictviolence/publication/ supporting-refugees-and-internally-displaced-through-development.

[13]Anyangwe, Eliza, 'Is it time to rethink the divide between humanitarian and development funding?' (2015) *The Guardian*, Accessed July 2018: https://www.theguardian.com/global-development-professionals-network/2015/dec/04/funding-humanitarian-assistance-development-aid.

[14]Ibid.

People who were displaced 10, 20 or more years ago are still displaced. This protracted displacement is a scandal. Tens of millions of people are living in limbo, eking out an existence with small amounts of humanitarian aid, unable to go home or to start afresh by settling in a new location... For at least twenty-five years, the gap between relief and development has been lamented and initiatives to address the gap have proliferated. But displacement is still widely seen as a humanitarian – not a development – issue... IDPs need support to recover land or property, to restore livelihoods, to find jobs.[15]

1.3   Displaced Persons in Sub-saharan Africa

These findings and recommendations for humanitarian and development efforts prove important for Sub-Saharan Africa, especially in addressing its significant refugee and internally displaced populations and ensuring the region can meet the Sustainable Development Goals. According to a UN report, Africa had more than five million refugees and more than 11 million IDPs by the end of 2016 in addition to having more than 450,000 asylum seekers and one million stateless persons.[16] For a global comparison, the UN High Commissioner for Refugees (UNHCR) writes that "Sub-Saharan Africa hosts more than 26% of the world's refugee population."[17]

Many causes lead people to becoming refugees and IDPs, including "conflicts, civil unrest, environmental disasters, oppressive regimes and concomitant abuse of human rights."[18] According to Adepoju, these can include "inter-ethnic and inter-religious conflicts and communal land disputes, which are aggravated by environmental disasters including flooding, soil erosion, and droughts," which cause the internal displacement of people.[19]

Therefore, given the significance of the refugee and IDP problem in Sub-Saharan Africa, it remains pivotal for the region to not only address the immediate humanitarian needs of these groups but also to consider the medium to long-term developmental needs of refugees and IDPs, especially if these groups are going to become enablers and drivers of economic growth rather than resource drains for already cash-strapped governments and relief organisations. According to the UNHCR, "When displaced persons can no longer rely on known ways of generating income, they have difficulty to adjust to new markets, learn new skills, and

---

[15]Ferris, E., '10 Years After Humanitarian Reform: How Have Internally Displaced Persons Fared?' (2015) *Brookings Institution*, Accessed June 2018: https://www.brookings.edu/blog/up-front/2015/01/28/10-years-after-humanitarian-reform-how-have-internally-displaced-persons-fared/.

[16]United Nations, 'Assistance to refugees, returnees, and displaced persons in Africa,' (2017) Accessed July 2018: https://reliefweb.int/sites/reliefweb.int/files/resources/N1726487.pdf.

[17]United Nations High Commissioner for Refugees, 'Africa' (2018) Accessed July 2018: http://www.unhcr.org/en-us/africa.html.

[18]Adepoju, Aderanti, 'Migration Dynamics, Refugees, and International Displaced Persons in Africa' (2016) *United Nations*, Accessed July 2018: https://academicimpact.un.org/content/migration-dynamics-refugees-and-internally-displaced-persons-africa.

[19]Ibid.

fully integrate with surrounding communities."[20] A potential solution to this
challenge, as suggested by the World Bank and other development institutions, is to
"provide quality financial services before, during and after periods of humanitarian
crises can improve people's resilience and help sustain livelihoods."[21] The German
Global Partnership for Financial Inclusion supports this view when it writes,
"Options to safely store money, to build-up (small) savings or send and receive
money transfers, and to carry out everyday life transactions are vital for FDPs
[forcibly displaced persons]."[22]

2   Space-Based Financial Services for Displaced Persons

Given this, a space-based financial service that stores and transacts wealth for these
types of populations could prove useful in alleviating the transition of these individuals from displacement to resettlement. The banking of individuals in
conflict-prone areas via satellites could allow them around-the-clock access to their
assets, while ensuring their assets are secured and easily accessible whenever and
wherever they may require them. This financial service could be offered at any stage
of a conflict-affected person's transition (pre-conflict; emigration; temporary settlement; post-settlement in either a new location or return to original location).
     Nonetheless, the primary objective of this concept would be to ensure the
safekeeping of affected populations' financial resources pre, mid, and post-conflict
in order to facilitate their transition and mitigate catastrophic financial loss affecting
their reintegration in society. The goal would be portability of assets for displaced
persons. National governments, international and intergovernmental development
organisations, and traditional financial institutions could prove as initial partners
and financers of such an initiative deploying space applications for the benefit of
Sub-Saharan Africa, its economies, and its displaced populations.

2.1   Unbanked Populations in Sub-saharan Africa

In addition to deploying satellites for the provision of financial services to assist
displaced persons, this could be done irrespective of an ongoing or current conflict
or crisis that results in displaced persons. These space-based financial services

---

[20]United Nations High Commissioner for Refugees, 'Action Sheet 16: Livelihoods' (2018)
Accessed July 2018: http://www.unhcr.org/4794b5f12.pdf.
[21]Pazarbasioglu, Ceyla, 'Financial inclusion for displaced people yields societal and economic
benefits for all' (2017) *World Bank*, Accessed July 2018: https://blogs.worldbank.org/voices/
financial-inclusion-displaced-people-yields-societal-and-economic-benefits-all.
[22]Global Partnership for Financial Inclusion, 'Financial Inclusion of Forcible Displaced Persons'
(2017) *Federal Ministry for Economic Cooperation and Development*, Accessed July 2018:
https://www.g20.org/sites/default/files/documentos_producidos/policy_paper_on_financial_inclus
ion_of_forcibly_displaced_persons_-_priorities_for_g20_action_gpfi_0.pdf.

could be implemented by development stakeholders proactively, helping to mitigate the consequences of displacement if and when it does occur, prior to its occurrence. This would not only support the humanitarian and development aim of aiding the transition of people from displacement to resettlement, but also address the high levels of unbanked people living in many of the refugee and IDP-affected countries within Sub-Saharan Africa. For example, as of 2016, Ethiopia had 659,000 refugees; Kenya had 551,000 refugees; Chad had 453,000 refugees; Uganda had 386,000 refugees, Cameroon had 264,000 refugees; and South Sudan had 248,000 refugees.[23] Respectively, these countries had the following levels of adults as an unbanked percentage of their population: 78% for Ethiopia, 25% for Kenya, 88% for Chad, 56% for Uganda, and 88% for Cameroon.[24] No data exists for South Sudan.

Conversely, countries that had the highest levels of IDPs in Sub-Saharan Africa, including Nigeria, Democratic Republic of Congo, and Sudan, had an unbanked populations at 56, 83, and 85%, respectively, as a proportion of their adult populations.[25] These figures help demonstrate that irrespective of the presence of a conflict or crisis producing displaced persons in any of these countries, significant majorities of these populations, except for Kenya, are unbanked and could benefit from space-based financial services, and which could pre-emptively address the portability of assets for displaced persons in the event a conflict or crisis does take place.

2.2 Cryptocurrencies for Storing and Transacting Wealth via Satellites[26]

These types of space-based financial services can be enabled through the use of cryptocurrencies. The system would rely on blockchain technologies in the form of cryptocurrencies to transact and store wealth. As system users will be operating internationally (ex. potentially displaced from country X to country Y and settled in country Z), it would be important for assets to be stored in a virtual and flexible currency that can be easily converted outside of traditional foreign exchange transactions. It is expected this system would be supported by a either a direct customer-to-satellite model or a customer-to-agent-to-satellite model. It is likely a mixed approach employing both of these models would be used given the varying contexts where conflicts are prone to take place leading to displaced populations.

---

[23]Adepoju, Aderanti, 'Migration Dynamics, Refugees, and International Displaced Persons in Africa' (2016) *United Nations*, Accessed July 2018: https://academicimpact.un.org/content/migration-dynamics-refugees-and-internally-displaced-persons-africa.

[24]World Bank, 'Universal Financial Access 2020' (2018) Accessed July 2018: http://ufa.worldbank.org/country-progress/.

[25]Ibid.

[26]Lindgren, David, 'Global Remittances and Space-based Cryptocurrencies: A Transformational Opportunity for the Post-2030 Agenda' (2018) *European Space Policy Institute*.

The growth in popularity of cryptocurrencies, based on blockchain technology, in recent years has attracted significant investment to that industry as individuals and companies seek more secure, cost-effective, and accessible means for their transactions. Companies now seek to host cryptocurrency networks on satellites in order to increase their security and accessibility.

For example, SolarCoin Foundation entered into a deal with a satellite company, Cloud Constellation, to "purchase data center capacity in space in order to secure its blockchain wallets from hacking."[27] Another company, Blockstream, announced in 2017 that it would launch a satellite network which would broadcast bitcoin, a type of cryptocurrency, around the world.[28] According to the company, "This will make the cryptocurrency more accessible to almost anyone, even in places where data costs are high and living standards and incomes are low."[29] Blockstream continues to write about its network: "The satellite network provides an opportunity for nearly 4 billion people without internet access to utilize bitcoin while simultaneously ensuring bitcoin us is not interrupted due to network interruption."[30] Finally, and most recently, a company that developed its own cryptocurrency, Nexus, has announced plans with a nanosatellite launch company, Vector, to host its block-chain across several satellites by 2019.[31] According to Nexus, "By using a satellite virtualization platform through GalacticSky [Vector's satellite-based software platform], Nexus can distribute its blockchain across multiple satellites, providing it enhanced reliability and performance. Nexus' secure cryptocurrency and decen-tralized peer-to-peer network will grant greater freedom and transactional trans-parency for global access to financial services."[32] The benefit hosting the cryptocurrency in space comes from it existing outside of an individual country and its ability to "create the backbone for a more decentralized financial ecosystem."[33]

As suggested by the review of the above described entrants into the space-based cryptocurrency sector, the space applications and space-based services sector retains significant potential for growth, particularly as these initiatives have only developed in the preceding two years as concepts with full deployment, at least in the case of Nexus, expected by 2019. These current innovations and those to come arise at a time when a confluence of trends emerge in the space sector, namely the introduction of cost-effective launch service providers, developments in technology

---

[27]Gallagher, Sean, 'Satellite cloud startup inks deal for space-based cryptocurrency platform' (2016) *Ars Technica*, Accessed January 2018: https://arstechnica.com/information-technology/2016/09/satellite-cloud-startup-inks-deal-for-space-based-cryptocurrency-platform/.

[28]Lant, Karla 'Bitcoin now comes from satellites in space. Welcome to the future' (2017) *Futurism*, Accessed January 2018: https://futurism.com/bitcoin-now-comes-from-satellites-in-space-welcome-to-the-future/.

[29]Ibid.

[30]Blockstream, 'Blockstream Satellite' (2017) Accessed January 2018: https://blockstream.com/satellite/blockstream-satellite/.

[31]Vector, 'Vector and Nexus Team Up to Bring Cryptocurrency to Space' (2017) *PRNewsWire*, Accessed January 2018: https://www.prnewswire.com/news-releases/vector-and-nexus-team-up-to-bring-cryptocurrency-to-space-300573678.html.

[32]Ibid.

[33]Ibid.

and greater accessibility of nano and microsatellites, and increased acceptance of cryptocurrencies and blockchain technology as a means of transacting and storing value.

2.3   Access Considerations for End Users

To facilitate access to the system, end users could have a mix of options for transacting and receiving information, including either a smartphone application made available across multiple mobile software platforms that transmits and receives data for facilitating transactions or a specifically designed and system-specific device for facilitating transactions. While mobile phone use, including smartphone use, has steadily increased among developing countries, its reach is limited to the most-underserved and least-developed areas where conflicts most occur, thus necessitating a system-specific device, likely utilizing affordable Raspberry Pi technology, for deployment to these locations. A recent International Telecommunications Union report confirms that a significant portion of least-developed countries' populations, where a great number of refugees and IDPs within Sub-Saharan Africa reside, still lack access to mobile phones, much less access to advance smartphone technology, when it stated that "By the end of 2017, the number of mobile-cellular subscriptions in LDCs (least developed countries) had increased to about 700 million, with a penetration of 70%."[34] This suggests that 30% of LDCs' populations lack access to mobile phones, let alone smartphones, despite LDCs having "launched 3G services and over 60% of their population are covered by a 3G network."[35] Therefore, an alternative option for interacting with the system would be necessary, made possible by an affordable technology such as Raspberry Pi that has been used in other development contexts.[36]

This access consideration for end users portends one area for further exploration as it would require the advent of a physical infrastructure comprising system-specific hardware that could facilitate transactions with the space-based system. Given the infrastructure that often accompanies groups of refugees and IDPs in the form of settlements and other humanitarian forms of assistance, such hardware could be deployed to those working in these locations, who could serve as agents and intermediaries between the consumers and satellite-based financial service.

2.4   Regulatory Challenges Associated with Cryptocurrencies

While some have embraced cryptocurrencies and the technology facilitating their use has proliferated, blockchain-based, digital currencies still face a number of

---

[34]CommsMEA, 'Mobile penetration reaches 70% in least developed countries of the world' (2018) *Arabian Industry*, Accessed June 2018: http://www.arabianindustry.com/comms/news/2018/jan/25/mobile-penetration-reaches-70-in-least-developed-countries-of-the-world-5876299/.
[35]Ibid.
[36]Raspberry Pi, 'About Us' (2018) Accessed June 2018: https://www.raspberrypi.org/about/.

challenges and hurdles that have emerged with the growth in their acceptance. In particular, these relate to the need to comply with anti-money laundering (AML) and know your customer (KYC) best practices. Financial regulators, such as the Securities and Exchange Commission (SEC) and Financial Industry Regulatory Authority (FINRA) in the United States, help ensure financial institutions and financial service providers comply with these practices as way of enforcing rules, regulations, and laws around banking, securities trade, and other financial transactions. Additionally, the Financial Crimes Enforcement Network (FINCEN), housed within the United States Department of Treasury, supports these efforts to ensure financial institutions do not facilitate transactions for nefarious purposes. One such law that serves as a global benchmark for banking and financial services is the USA PATRIOT Act, a law established in the United States following the 11 September 2011 terrorist attacks. FINCEN summarises the USA Patriot Act in the following ways:

> To strengthen U.S. measures to prevent, detect and prosecute international money laundering and financing of terrorism; To subject to special scrutiny foreign jurisdictions, foreign financial institutions, and classes of international transactions or types of accounts that are susceptible to criminal abuse; To require all appropriate elements of the financial services industry to report potential money laundering; To strengthen measures to prevent use of the U.S. financial system for personal gain by corrupt foreign officials and facilitate repatriation of stolen assets to the citizens of countries to whom such assets belong.[37]

According to Iza Wojciechowska, "KYC laws were introduced in 2001 as part of the Patriot Act, which was passed after 9/11 to provide a variety of means to deter terrorist behavior. The section of the Act that pertained specifically to financial transactions added requirements and enforcement policies to the Bank Secrecy Act of 1970 that had thus far regulated banks and other institutions. These changes had been in the works for years before 9/11, but the terrorist attacks finally provided the political momentum needed to enact them."[38] Thus, since 2001, the United States financial services industry, and the global financial services at large, have moved toward increased efforts to mitigate illicit financial transactions, particularly those in service of terrorism. Similarly, according to McKinsey analysts as a response to the 2007-2008 financial crisis, "Compliance risk has become one of the most significant ongoing concerns for financial-institution executives. Since 2009, regulatory fees have dramatically increased relative to banks' earnings and credit losses... Additionally, the scope of regulatory focus continues to expand. Mortgage servicing was a learning opportunity for the US regulators that, following the crisis, resulted in increasingly tight scrutiny across many other areas (for example, mortgage fulfillment, deposits, and cards). New topics continue to emerge, such as conduct risk, next-generation Bank Secrecy Act and Anti-Money Laundering

---

[37]Financial Crimes Enforcement Network, 'USA PATRIOT Act' (2018) Accessed July 2018: https://www.fincen.gov/resources/statutes-regulations/usa-patriot-act.
[38]Wojciechowska, Iza, 'What is KYC and why does it matter?' (2018) Accessed July 2018: https://fin.plaid.com/articles/kyc-basics.

(BSA/AML) risk, risk culture, and third- and fourth-party (that is, subcontractors) risk, among others."[39]

In the context of these two major events, the terrorist attacks of 2001 and the global financial crisis during 2007-2008, compliance, particularly around AML and KYC best practices, has been at the forefront of the financial industry. Therefore, cryptocurrencies, which offer users degrees of anonymity, pose unique compliance challenges in order to comply with the financial industry's best practices, and any financial service provider operating in the cryptocurrency sector, including those transacting via space-based satellites, would need to ensure those transacting using via its services are identifiable and low-risk users of the service. This could prove especially challenging for refugees' and IDPs' circumstances where identification and known records of individuals may not exist or be readily accessible; however, the introduction of a model of operations that features the use of agents in refugee/IDP settlements could mitigate risks in this regard as the agents could be the responsible parties for ensuring the legitimacy of the users and their transactions. Receipts and physical documentation provided by the agent would need to be provided to the consumer in order to facilitate their access to the system at a later time, either be it with the same agent in the same settlement or upon resettlement and engagement with a different agent.

3   Financing and Governance of a Space-Based Financial Service Provider

As a space-based financial services provider primarily serving displaced populations in Sub-Saharan Africa and helping efforts to achieve SDGs 8 and 10, global development partners could prove the most viable source of financing and initial support to such a concept. These could include national governments where conflict-affected areas exist, bilateral and multilateral development organisations, and traditional financial institutions. A service as described here would be in the interests of the global development community, and thus early supporters include these various stakeholders. While development of an entirely new satellite constellation would likely prove unnecessary so long as capacity on existing space-based systems existed to transact and store the information require of a cryptocurrency financial service provider, humanitarian and development organisations would prove as instrumental early partners in the piloting and development of the programme, and could include UN agencies and large bilateral donor organisations such as the United States Agency for International Development (USAID), the European Union (EU), the United Kingdom's Department of International Development (DFID), and the Swedish International Development Cooperation Agency (SIDA) among others.

---

[39]Kaminski, Piotr and Robu, Kate, 'A best-practice model for bank compliance' (2016) *McKinsey&Company*, Accessed July 2018: https://www.mckinsey.com/business-functions/risk/our-insights/a-best-practice-model-for-bank-compliance.

Development organisations and the public sector agencies of conflict-affected countries, including of national and international governmental bodies, could be enticed to support the concept given its immediacy and relevance in responding to a pressing humanitarian and development need. Growing levels of conflict leading to an ever-increasing number of displaced persons, including IDPs and refugees, are creating unstable and unsustainable levels of need to support these populations for not only their immediate term displacement but also for their medium to long-term resettlement and reintegration.

A space-based financial service for refugees and IDPs could also prove enticing to traditional financial institutions and investment groups given the potential for the system to evolve with expanded financial services beyond transacting and savings, which could include lending and insurance and investment. The short to medium-term goal of this system would only feature servicing the wealth trans-acting and storage needs of refugee and IDP populations in order to alleviate and improve their transition from displacement to resettlement, whereas a long-term goal could include theirs and other populations' more significant financial inclusion in a country's banking and overall economic system.

4   Conclusion

A space-based financial service dedicated to serving vulnerable, displaced populations could yield significant humanitarian and development-oriented results in the short to medium term, while demonstrating long-term viability and sustainability as its suite of space-based banking services grew. This could be made possible using a constellation of multiple satellites providing around-the-clock global coverage, allowing users to store and transact wealth using blockchain technologies, namely cryptocurrencies, which would enable their access whenever and wherever they required their assets. A space-based system such as this would remove the catastrophic implications of displaced persons losing their assets during conflict and other humanitarian situations, and offer a smoother transition from displacement to resettlement.

Author Biography

David Lindgren is an experienced international development professional with a strong knowledge and practice in program development and management. David joined as Assistant Director of the Research Office at American University's School of International Service, where he works closely with faculty and doctoral students throughout the project development process from idea to submission. Previously, David served as a program officer with Freedom House in Johannesburg, South Africa, where he implemented democracy and human rights programs. Prior to this, he worked in Washington, DC supporting programs across Central, West, and Southern Africa. David graduated from American University's School of International Service in 2011 with a B.A. in International Studies, and is currently a candidate for an M.Phil. in Space Studies at the University of Cape Town.

The Importance of Internet Accessibility and Smart City in Sub-Sahara African Region Through Space Technology

Tivere Hugbo

Abstract

This observation correlates the importance of Internet accessibility and the Smart City concept in Sub-Sahara African, through space technology. The Internet has foster innovation through peer-to-peer networking, providing ease to data communication networks, e.g. electronic mail, & E-education. It has become part of our daily utility with very high adoption rate. Europe for example has more than 85% of her household using the internet as at 2016, while that of Africa was around 21% (The Statistic Portal. Website: https://www.statista.com/topics/3853/internet-usage-in-europe/). This disparity is mainly due to inadequate technological infrastructures. Smart city is built on Information and Communication Technology (ICT), which most countries in Africa lacks. With robust internet connectivity, Africa can close the gap by tapping into the enormous resources provided by the internet, to support the smart city approach which can leapfrog Africa to the 21st century in terms of technology and Innovation. Space technology such as the OneWeb constellation of small satellites, will be launched to the Lower Earth Orbit, providing Global internet broadband services as early as 2019. Such access to the internet can provide solutions to the Smart City concept in Africa as well as supporting the United Nations Sustainable Development Goals.

T. Hugbo (✉)
Space Studies Department, University of Cape Town, Cape Town, South Africa
e-mail: Tiveforme@gmail.com

T. Hugbo
Mathematics and Economics, University of Benin, Benin, Nigeria

© Springer Nature Switzerland AG 2019
A. Froehlich (ed.), *Embedding Space in African Society*, Southern Space Studies,
https://doi.org/10.1007/978-3-030-06040-4_8

105

1 Introduction

The United Nations Sustainable Development Goals was set to combat poverty, environmental protection and prosperity for all.

Infrastructure plays a huge role in economic development by providing employment, innovation and constructive competition. Basic infrastructures like water irrigation, road networks, energy power provision and educational institutions increase life expectancy and provide multiple sources of employments. The manufacturing industries accounted for over 16% of the world workforce as at 2009 and about 20–23% in 2013. Investing in Industry, innovation and Infrastructure in Africa, will not only promote her well-being, but provide basic human needs, like clean water, robust road network, and adequate health care, the list is endless. Investing in the technology Industry alone will affect the whole chain of development, i.e. security, urbanization, agriculture, good governance, and high standard of living.

The impact of internet connectivity in Africa have seen progress over time in different industries. In 2010, the Global Information and Communication Technologies (GICT) Department study, shows that between 2000 and 2006 developing countries broadband penetration brought about 1.38% point for every 10% increase in broadband penetration.[1] This analysis entails that broadband penetration in emerging countries like South Africa, Nigeria, Kenya, etc. provides higher Gross Domestic Product than that of developed countries which realize about 1.21%-point increase in GDP with each 10% increase in broadband penetration. Innovation through Information Communication Technology (ICT), has seen significant growth in a country like Kenya and this can be attributed to the high level of internet penetration, with internet speed as high as 13.7 Mbps,[2] which is the highest in Africa, country like Kenya is setting the pace for Tech start-ups and projects in Africa presently, followed by South Africa with the second highest rating in terms of internet connectivity, speed and access. What this means is with Internet penetration, technological innovation is eminent, and will cause a positive change in other parts of a developing countries.

2 Internet Access in Africa

Though there has been progress in making Africa more connected, the International Telecommunication Union (ITU) estimated that only 21.8% of the African population had access to the internet and 40% in few selected Africa countries as at

---

[1]Worldbank.org. (2018). *Digital Dividends*. [online] Available at: http://pubdocs.worldbank.org/en/391452529895999/WDR16-BP-Exploring-the-Relationship-between-Broadband-and-Economic-Growth-Minges.pdf [Accessed 9 May 2018].
[2]Medium. (2018). *Internet Access in Kenya—DataScience LTD—Medium*. [online] Available at: https://medium.com/@DataSciencing/internet-access-in-kenya-b7063700a49f [Accessed 10 Apr. 2018].

2017, we are still 20 years behind developing continents like Europe or the Asia pacific in terms of internet usage and penetration. This statistic is alarming and has dwindled and still affecting technological innovation. Though, we are in the age of free big data via the Internet, most African countries are still yet to be exposed to such access, and as we know Information expands knowledge, and with a higher penetration of the internet, Africa can tap into the vast resource of information therefore enhancing economical and Industrial growth.

Factors that affects the growth of technological innovation are labor shortage, technological know-how, rigid policies, tax regimes, and high cost of implementation. Since most connectivity in past years was through fiber Connections it is still very expensive to get connected. In recent times the cost of internet connection has seen some reduction and this is due to satellite connectivity and mobile internet broadband connectivity, but it's still not cost effective enough to allow higher penetration in the African context.

3 Sustainable Cities and Communities

Building sustainable and safer cities, increase quality of life, create an echo-friendly environment while monitoring climate change, and provide affordable housing with basic services. Africa has the highest rate of population growth within the last decades, growing at a pace of 2.55% annually between 2010 and 2015, it will account for more than half of the world's population in 2050, according to the United Nations population census. Africa will play a central role in population distribution in the coming decades. The present estimated population in Africa is about 1.3 Billion, comparing with the world population of 7 Billion.[3] Reasons for sporadic growth in Africa population are associated with factors like: High fertility rate, increase in average life span, Child mortality rate and more, even with an uncertain fertility rate, Africa is still estimated to grow in population due to its high percentage of younger population of about 60%. Life expectancy in Africa has grown from 2 years in 1990s to 6 years in the 2000s.

4 Urbanization Trends in Africa

Urbanization mounts pressure on public infrastructure in Africa, such as increase demands for jobs, energy, housing, clean water, food, transportation infrastructure, social services, etc. Urbanization in Africa has digressed into illegal settlements and slums over the past decades; The North of Africa has the highest Urbanization rate of 47.8%, while Sub Saharan Africa covers about 32.8% but higher slum settlement

---

[3]United Nations. (2018). Population Census. [online] Available at: http://www.un.org/en/sections/issues-depth/population/.

of about 65%. Reasons for the high disparity of slums between these regions are based on issues like, but not limited to lack of basic infrastructure. As at 2010 only 20% of sub-Sahara Africa had access to electricity, 3% had access to fixed network broadband and 53% had access to mobile phones, 84% have access to clean water and 54% have access to sanitation. About 60% of African citizens do not have access to clean water and sanitation, as reported by the World Energy Outlook (WEO), through the International Energy Agency (IEA) in 2016.

With the world total population growth dropping from 1.24 to 1.18% within the past 10 years, Africa population seems to be increasing. Africa is tending towards a demographic inflection point, at this point the steady growth in population might be of significant advantage or disadvantage. Urban settlement will see a sharp rise by over 300 million between 2000 and 2030, Nigeria, Democratic republic of Congo, Uganda, Ethiopia, and United republic of Tanzania are amongst the countries in the world that are expected to have high concentration of population growth between 2016 and 2050 (UN Population Census, 2016). This is due to their past and present contribution to the world population.

Urbanization in Africa will have its biggest shifts in the coming decades; by 2030 six of the world's mega cities will be situated in Africa, cities like, Cairo, Lagos, Greater Johannesburg, Nairobi and Kinshasa-Brazzaville, will see a significant rise in population, given that these cities are huge part of their countries economic capabilities.[4] By 2030, cities like Lagos, Kinshasa and Cairo will be housing more than 20 million people, while Johannesburg, Nairobi and Kinshasa-Brazzaville will be well above 10 Million people. These levels of population growth will *ceteris paribus* increase demands for better urbanization like, real estate, public services and industrial infrastructure.

5 Smart City and Sub-sahara Africa

Smart cities can be the bridge with which Africa can leapfrog her technological and innovative drag. A smart city is an "innovative city that uses information and communication technology (ICT) and other means to improve quality of life, efficiency of urban operations and services, and competitiveness, while ensuring that it meets the needs of present and future generations with respect to economic, social environmental aspects".[5] As defined in a paper released by the United Nations Economic and Social council in 2016.

There have been numerous definitions of a smart city over time but in this context, we will define a smart city as one that uses Information and Communication Technology (ICT), such as the Internet of Things (IoT), Big Data, sensors, etc.

---

[4]Institute for Security Studies. Julia Bello-Schünemann, Senior Researcher and Ciara Aucoin, Researcher, African Futures & Innovation. Website: https://issafrica.org/iss-today/africas-future-is-urban.
[5]Unctad.org. (2018). Smart City. [online] Available at: http://unctad.org/meetings/en/SessionalDocuments/ecn162016d2_en.pdf [Accessed 2 Jun. 2018].

to improve the lives and well-being of her people and its environment. Examples of components that can be labelled smart are but not limited to, roads & traffic monitoring, police alert systems in case of a robbery, waste management through sensors and the IoT, well organized and planned urbanization development, through satellite imaging and drones, weather monitoring in real time, Block-chain technology for security and transparency, the lists are almost endless.

Nairobi and Cape Town are the leaders of the smart city concepts in Africa, with Nairobi the capital of Kenya being at the forefront, they have seen significant growth in urban development and infrastructure. Nairobi has been awarded[6] *"The most Intelligent city"* in Africa two years in a roll by the Intelligent Community Forum, a research institute dedicated to how ICT can tackle social and governmental challenges to improve lives in Urban areas.

Konza Tachno City is already deemed the Silicon Savannah in Kenya, this city is 60 km from the center of Nairobi, this satellite city is still undergoing construction. The city will be embedded with sensors across buildings, roads, transportation systems, etc. which will relay and gather data to be used for optimization of traffic controls, security management, sustainable housing and resource control, while involving the participation of the civic society. These improvements in ICT and smart initiatives is connected to the mobile broadband connectivity that has penetrated Kenya as a whole, *"How we made it in Africa"* published an article in 2017, stating that Kenya saw the increase of mobile phone subscriptions per 100 persons from 13.5 to 81.9 between 2005 and 2015. Based on the network coverage provided by these mobile broadband, small and large-scale businesses are seeing more opportunities to grow faster and sustainably.

Kenya has become a world leader on the use of mobile phones for money transfer and other services, which in turn has filled a niche for local banks to initiate less cost in money transfer fees. Example of such applications is the *"M-Pesa"*, this application allows people in Kenya to make money transactions and pay bills seamlessly, *M-Pesa* recorded about US$28B in transactions in 2015 alone. *Safaricom*[7] one of the leading company in Kenya, providing services from charity foundations to mobile broadband distribution, Apps. creation and economic empowerment, has reported enabling over 1 million children to access interactive and educational materials through mobile phones, they have also created *"My data manager"* an application used by costumers to manage their data consumptions. More so they are adhering to the SDG goals by adopting 9 out of the 17 goals, which has they stated, is one of the core foundation of their businesses, in giving back to the people of Kenya and humanity as a whole.

---

[6]Veras, O. (2018). Smart cities in Africa: *Nairobi and Cape Town*. [online] How We Made It in Africa. Available at: https://www.howwemadeitinafrica.com/smart-cities-africa-nairobi-cape-town/58209/ [Accessed 15 May. 2018].

[7]Safaricom. (2018). *Safaricom Annual Report & Financial Statement*. [online] Available at: https://www.safaricom.co.ke/gonext/pdf/Safaricom_Annual_Report_2017.pdf [Accessed 5 Jun. 2018].

South Africa is one of the countries in the Sub-Sahara Africa region that is taking the Smart City approach. Cape Town for example has be labelled one of the country's smartest cities, the city launched a four-pillar project namely: Digital infrastructure, Digital inclusion, E-governance and Digital economy. The city of Cape Town has an "*Open Data Portal*" through which registered citizen of the city can find their data publicly when searched for, more so IBM launched a Fire Management Portal, that uses data from the city's open data platform to better manage and predict potential fire hazards in real time, increasing emergency fire responses. The country of South Africa has also employ using remote utilities meter reading, through the internet of Things (IoT) to eliminate manual data-capture errors and support with water managing water usage and energy consumption by 10%.[8]

5.1 Space Technology and Smart City in Africa

Space technology can support the accessibility of the internet which in turn enhances the growth of technology innovation through the smart city concept. The wide adaptation of the mobile broadband internet connectivity in the region indicates high demand for internet access.

With the proposed innovative project been carried out by the space Industries pioneers like, *OneWeb*, SpaceX and other tech power houses, we can expect a broader penetration of the internet, hence promoting sustainable development through technology innovation. *OneWeb* is initiating a fleet of satellites that will deliver affordable internet access globally, with these small and cheaper satellites (CubeSats) roving around the Low Earth Orbit with peak speed of 500 mbps as their initial roll out and then increases to 2.5 gbps on their second constellation phase and being in the lower earth orbit will support with the high latency problems as well as affordable self-installed terminals. *Oneweb* which partners with Airbus one of the leaders in the space industry will be able to provide cheaper and easily accessible internet usage. They have a planned launch of 900 satellites in the low earth orbit by 2018, they proposed that by 2027 they will bridge the world's digital divide and by 2022 connecting every unconnected schools and more.

Africa can benefit from the *OneWeb* constellation plans by piggy bagging and taking advantage of the provided services of such project, which will be cost effective. Since the adaptation of mobile broadband is already widely spread and has crated wide range of opportunities, the *oneWeb* constellation project will provide even more wider coverage to remote areas that has never been introduced or have access to the internet before. What this brings is a revolution that we are already experiencing in Kenya and other parts of Africa, this can be compared to the Mobile phone revolution in Africa during the early 2000s. The mobile phone

---

[8]Outside Insight. (2018). Smart cities Africa: Cape Town and Nairobi take the lead—Outside Insight. [online] Available at: https://outsideinsight.com/insights/smart-cities-africa-cape-town-nairobi-take-lead/ [Accessed 26 Jun. 2018].

technology bridged the gap between telecommunication accessibility in Africa by bypassing the traditional fixed line networks that was not evenly distributed across the region. With the provision of accessible internet, the smart city concept can easily come to fruition, since the Internet is one of the major bedrocks of a smart city and technology innovation, there is no doubt that these constellations proposed by *Oneweb, SpaceX* and *Google*, will transform the Innovative space across different industry in Africa.

6 Conclusion

In conclusion, this observation creates a bridge between, Internet penetration, technology innovation and the smart city concept in Africa. It also shows how connecting these entities can bring about solving these United Nations Sustainable Development Goals, by making our Urban space more resilient, sustainable, secure and inhabitable, hence bringing technological innovation and infrastructure.

With the implementation of a steady and consistent internet connectivity in Sub-Sahara Africa, the region can have access to unlimited resources and information that will elevate her present dysfunctional state.

Author Biography

Tivere Hugbo is from Nigeria and received his Diploma in computer engineering, B.Sc. in Mathematics and Economics, and he is currently pursuing an Mphil. in space studies through the University of Cape Town, Space Lab. His dissertation topic is focused on using Space technology as an enabler to a smart Africa.

Aerospace Research in African Higher Education

Christine Müller

Abstract

In recent years, some African countries have tried to highlight and exploit the potential of Space Sciences. But many countries on the African continent have far greater problems than investing in a very cost-intensive future technology: population growth, insecurity and conflicts, insufficient infrastructure etc.. However, with enthusiasm for outer space, interest in science, mathematics and technology is also on the rise. From this point of view sound school education, the creation of training paths in aerospace technology, study programs on aerospace science and research collaboration in this area are important components of sustainable development. Aspects that stand out are 1. Why Africa must invest in space research, 2. Some African countries already recognize and appreciate the value of aerospace research. 3. Partnerships— South-South as well North-South are crucial for successful programs. A description of current trends in selected countries with space programs or recognizable efforts in aerospace research is summarized according to strategies, concepts and ideas leading in this direction.

In recent years, some African countries have tried to highlight and exploit the potential of Space Sciences. But many countries on the African continent have far greater problems than investing in a very cost-intensive future technology: population growth, insecurity and conflicts, insufficient infrastructure etc. However, with enthusiasm for outer space, interest in science, mathematics and technology is also on the rise. From this point of view sound school education, the creation of training paths in aerospace technology, study programs on aerospace science and research collaboration in this area are important components of sustainable development.

C. Müller (✉)
International Office, Head of Department "European and International Networks",
University of Bonn, Bonn, Germany
e-mail: christine.mueller@uni-bonn.de

© Springer Nature Switzerland AG 2019   113
A. Froehlich (ed.), *Embedding Space in African Society*, Southern Space Studies,
https://doi.org/10.1007/978-3-030-06040-4_9

As the Sustainable Development Goals (SDGs) of the 2030 Agenda for Sustainable Development came officially in 2016 into force,[1] this paper aims at pointing out how efforts in space embedment can support African countries to realize at least four of these SDGs:

Goal 4. Ensure inclusive and equitable quality education and promote lifelong learning opportunities for all;
Goal 8. Promote sustained, inclusive and sustainable economic growth, full and productive employment and decent work for all;
Goal 9. Build resilient infrastructure, promote inclusive and sustainable industrialization and foster innovation;
Goal 17. Strengthen the means of implementation and revitalize Global Partnership for Sustainable Development.

Aspects that stand out are 1. Why Africa must invest in space research, 2. Some African countries already recognize and appreciate the value of aerospace research. 3. Partnerships—South-South as well North-South are crucial for successful programs. A description of current trends in selected countries with space programs or recognizable efforts in aerospace research is summarized according to strategies, concepts and ideas leading in this direction.

The term "Aerospace Research" contains in the context of this article all scientific disciplines that are involved in the exploration of aerospace: aeronautics, space system, energy, transport, materials, digitalization and security. The scientific and social value of exploring aerospace, solar system and meteorites is widely undisputable. Equally clear are the benefits of this science for sustainable development, innovation and economic growth like the ability to monitor climate change, measure pollution and talk on cellular phones. So points the African Union's strategy paper about African Space Policy out, that space services in the United Kingdom generate £ 7 billion annually and created over 70.000 jobs.[2] Furthermore, research in this field often includes physicists, chemists, geologists, biologists and engineers in multidisciplinary groups. This multidisciplinary approach makes it the ideal model for solving challenges of our times such as climate change and food security given that the complexity of such issues cannot be solved by a single person as representatives of a singular discipline. High-quality education in science, technology, engineering and mathematics (STEM) would help to empower communities to improve and directly address their specific problems. The embedment of

---

[1]United Nations General Assembly: Transforming our world: the 2030 Agenda for Sustainable Development (A/RES/70/1). The term "Sustainable Development" has been defined by World Commission on Environment and Development (WCED) in 1987 as "the development that meets the needs of the present without compromising the ability of future generation to meet their own needs". Our Common Future. Chapter 2: Towards Sustainable Development, WCED/A/42/427.
[2]African Union: African Space Policy: Towards Social, Political and Economic Integration: https://au.int/sites/default/files/newsevents/workingdocuments/33178-wd-african_space_policy_-_st20444_e_original.pdf, June 16, 2018, page 5 (extracted from Satellite and Space Services – Intellect Technology Association, UK, Intellect Publication 2013).

space to African Society could be used as a push-factor to attract more pupils and students to STEM subjects, which would in turn help to "build resilient infrastructure, promote inclusive and sustainable industrialization and foster innovation" (SDG 9).

In recent years, some African countries have tried to highlight and exploit the potential of Space Sciences. However, the overall impression is that low-income and lower-middle income economies in Sub-Saharan Africa focus on their most urgent daily problems when it comes to allocation and distribution of government expenditure. Under these circumstances, funding basic scientific research cannot be expected to be on the top of the priorities list. In short summary and condensed statement: When it comes to education in Africa, it is not possible to compare one state to the other, and even within one state, there are differences between rural and urban areas and between different zones and regions. Variety of languages, the lack of appropriate facilities and qualified educators, emigration and violent conflicts— only to mention a few of the most pressing problems of some of the low-income and lower-middle income economies pose serious challenges. Nevertheless, even under these conditions, it is a fundamental issue "to ensure inclusive and equitable quality education and promote lifelong learning opportunities for all" (SDG 4).

On the other hand, promising development can also be observed: The UNESCO Science Report, Towards 2030, points out that for sub-Saharan Africa, the years 2010–2015 were a period of strong economic growth, with more emphasis on Science, Technology and Innovation (STI). There are several positive examples in this area, although relatively little funding is still invested in Research and Development (R & D) in the region.[3] Yet several countries including Ethiopia, Ghana, Mozambique and Rwanda are experiencing strong growth in scientific production. "Although South Africa accounted for 46% of sub-Saharan African publications in 2014, low-income countries such as Benin and Gambia have comparable scientific productivity (per million people) to those of middle-income countries. Ethiopia (0.61% in 2013), Kenya (0.79% in 2010) and Mali (0.66% in 2010) have their R & D efforts (GERD [gross domestic expenditure on R&D] as % of GDP) in recent years raised to the level of a middle-income economy. Malawi even dedicates 1.06% of GDP to R & D, the highest rate in Africa; Scientists from this country publish more than any other country of similar size in the mainstream journals."[4] However African scientists are still underrepresented in Planetary and Space Sciences (PSS) as "a preliminary scan of articles published between 2000 and 2015 in four representative PSS journals—*Icarus, Journal of Geophysical Research: Planets, Journal of Geophysical Research: Space Physics*, and *Meteoritics and Planetary Science*—reveals that Africa produces less than 1% of the

---

[3]Although sub-Saharan Africa reached an additional 1 percentage point of the world's population (12.5%) between 2007 and 2013, its gross domestic product (GDP) grew by only 0.3% and gross domestic expenditure on R & D (GERD) by only 0.1%; see: https://en.unesco.org/unesco_science_report/africa; June 30, 2018.

[4]ibid.

world output of scientific publications in PSS, despite having more than 15% of the world's population."[5]

All in all, it can be said that African society started late to explore space; initiatives are rare and rather punctual. This article will attempt to give an overview of current projects and points out strategies and ideas for the future.

1 Previous Development Using the Example of Selected Countries (in Alphabetic Order)

1.1 Egypt

Although not within the Sub-Saharan Africa region, Egypt started its official space program in 1960. The National Authority for Remote Sensing & Space Sciences (NARSS) was established in 1991 and belongs to the Ministry of Scientific Research. Objectives are: Enabling Egypt to join Space Technology Age through designing and manufacturing small satellites; transfer of advanced space technologies in communication, computers, programs, optics, sensors, new materials, command and control and energy to the Egyptian Scientific community; Utilizing of space technologies and applications in development plans; Acquiring national capabilities in Space Technology disciplines; Establishment of scientific and industrial base in advanced technology fields; Building human resource capabilities for space science fields; Coordinating and enhancing the cooperation between the research and industrial centers and space program through a national program.[6] Nowadays there are plans by the Egyptian government to eventually establish a national space agency. An effort towards this was made in August 2016.

Since 2011, the Egyptian Ministry of Education has opened 9 STEM boarding schools, 10–12 grade with project-based learning groups, a MIT Fab Lab and strong collaboration with universities, research centers and companies. They want to educate students who are problem solvers, innovators, inventors, self-reliant, logical-thinkers and technologically literate.[7]

Egypt has an infrastructure in the field of aerospace: NARSS, the Planetarium Science Center—Bibliotheca Alexandria, a Space Generation Advisory Council, The Academy of Scientific Research and Technology (ASRT) on the one hand, and academic environment like the Cairo University Aerospace Engineering Department, Egyptian National Research Institute of Astronomy and Geophysics (NRIAG) and Zewail City of Science and Technology with their earth observation satellites program.

In 2018, Egypt will be hosting the African Association of Remote Sensing and the Environment (AARSE) Conference with the topic Earth Observation and

---

[5]The State of Planetary and Space Sciences in Africa, see: https://eos.org/features/the-state-of-planetary-and-space-sciences-in-Africa, June 4, 2018.
[6]www.narss.sci.eg, July 2, 2018.
[7]www.stemegypt.edu.eg, July 2, 2018.

Geospatial Science in Service of Sustainable Development Goals.[8] The conference is organized by AARSE and the Arab Academy of Science and Technology, in partnership with NARSS. The AARSE international conference is conducted biennially across Africa, and commenced in 1996, in Harare (Zimbabwe).

1.2 Ethiopia

By its geographical position and clear sky, Ethiopia is the ideal location for space observation. In 2015, the Ethiopian Space Science Society (ESSS) with its 10,000 members has opened the Entoto Observatory and Research Center (EORC) near its capital city Addis Ababa. The institution works with the motto "We explore the universe for the benefit of our people". The multi-million-dollar project originated at a time of greatest drought, when millions of people were at risk of famine. For the government, it was crucial to still invest in fundamental research: earth observation can be used to improve agriculture or lower communication costs: "The director explains how agriculture, water resources, telecommunications, education, health-care and the economy all stand to gain from Ethiopia's space science research. What may seem like a huge investment now will pay off in years to come, he says."[9]

It is remarkable, that the ESSS supports 60 Space Science Clubs at schools throughout the country. But on university level, the Addis Ababa Institute of Technology offers only 24 places in a PhD study course in Astronomy. Capacity building in this section would be desirable to support Ethiopia to reach the next step: Ethiopia announced in 2017, that it would launch a satellite into orbit within the next years to help monitor weather conditions.

1.3 Ghana

Ghana started the Second Science and Technology Policy in June 2012 with one of the objectives to strengthen capacity development in STI, but also opened the Ghana Space Science and Technology Centre (GSSTC) in 2012. It is partner in the Square Kilometre Array (see: South Africa): A communications antenna in Ghana has been converted from being a redundant telecoms instrument into a functioning very long baseline interferometry (VLBI) radio telescope. So, Ghana became the first partner country of the African VLBI network (AVN) to complete conversion of communications antenna into a functioning radio telescope. The 32 m converted telecommunications antenna at the Ghana Intelsat Satellite Earth Station at Kutunse

---

[8]12. AARSE International Conference: Earth Observations and Geospatial Science in service of Sustainable Development Goals; www.aarse2018.org.
[9]The Guardian: Natahsa Stallard: They call us crazy: a trip to Ethiopia's first space observatory, https://www.theguardian.com/world/2016/mar/21/ethiopia-first-space-observatory-drought, March 21, 2016.

will be integrated into the African VLBI Network (AVN) ahead of second phase construction of the SKA across the African continent.[10]

Although the governmental space program faces criticism in view of other national challenges, as mentioned by Chris Matthews,[11] Ghana has sent its first satellite GhanaSat-1 into orbit in 2017. The satellite was developed by students of the All Nations University in Koforidua and funded by the Japanese Space Agency (JAXA).[12]

1.4 Kenya

In 2013, the Science, Technology and Innovation Act[13] was passed by parliament, contributing to the realization of the Kenyan Vision 2030, which aims to "transform Kenya into a newly industrializing, middle-income country providing a high quality of life to all its citizens by 2030 in a clean and secure environment".[14] Under the Act a National Commission for Science, Technology and Innovation, an advisory board in charge for quality assurance, the National Innovation Agency and the National Research Fund were established. The act made provisions for the fund to receive 2% of Kenya's GDP each financial year. This substantial commitment of funds should enable Kenya to reach its target of raising GERD from 0.79% of GDP in 2010 to 2% by 2014.[15]

Kenya launched its first space satellite in 2018, designed by experts at the University of Nairobi in collaboration with the University of Rome under the KiboCUBE Programme. KiboCUBE is an initiative that offers educational and research institutions from developing countries, the opportunity to deploy cube satellites from Japanese Kibo module to the International Space Station (ISS). The Kenyan satellite was developed under the United Nations Office for Outer Space Affairs (UNOOSA) and Japan Aerospace Exploration Agency (JAXA). This dedicated collaboration is designed to improve space technology of the developing countries of the United Nations member states. In August 2016, the CubeSat was selected jointly by UNOOSA and JAXA for the first round of KiboCUBE program.[16] The satellite will be used to observe farming trends.

---

[10]www.ska.ac.za/media-releases/ghana-and-south-africa-celebrate-first-success-of-african-network-of-telescopes/, July 4, 2017.

[11]"Still, one of the biggest challenges for Ghana's space industry remains addressing criticism that it's irresponsible to spend government funds on space initiatives in a country where even amid great urban development, poverty still affects 20 percent of the population, a nationwide electricity crisis continues, and corruption is rampant." Matthews, Chris: Why Ghana started a Space program, https://motherboard.vice.com/en_us/article/nz7bnq/why-ghana-started-a-space-program, January 5, 2016.

[12]Ghana launches its first satellite into space, www.bbc.com/news/world-africa-40538471, July 7, 2017.

[13]www.education.go.ke.

[14]www.vision2030.go.ke.

[15]UNESCO report, page 523.

[16]www.global.jaxa.jp/press/2018/01/20180119_kibocube.html, May 25, 2018.

The deployment of the satellite into space was broadcasted live by Kenya Broadcasting Corporation (KBC), the state-run media organization. Interested public were invited by the University of Nairobi to follow the events live from Japan and the ISS.[17]

Besides newly invented aerospace research, Kenya is building capacity in the energy research area and in helping start-ups to capture markets through initiatives such as iHUB,[18] m:Lab East Africa[19] and @iLabAfrica, established in January 2011 as a research centre within the Faculty of Information Technology at Strathmore University, a private establishment based in Nairobi.[20] It furthers research, innovation and entrepreneurship in ICTs. Noticeable projects in renewable energy are the Olkaria Geothermal Power Plant, Rift Valley and the wind park Lake Turkana. Technology incubators like Nairobi Industrial and Technology Park and Konza Technology City 'Konza Technopolis' support the government on the way to SDG 9. Nairobi Industrial and Technology Park is being developed within a joint venture with Jomo Kenyatta University of Agriculture and Technology. Konza Technology City has been constructed since 2013 and in the first phase it is expected to create over 20.000 direct and indirect jobs.[21] Spending on R&D is on the rise in most countries with innovation hubs. Kenya now has one of Africa's highest R&D intensities (0.79% of GDP in 2010).[22]

1.5 Nigeria

The African Regional Centre for Space Science and Technology in English (ARCSSTE-E) in Nigeria was established in 1998 as one out of two regional centers in Africa. The other one is in Morocco (for the francophone countries). "The vision of ARCSSTE-E is the development of human resources and creation of public awareness on the benefits and applications of space science and technology for sustainable development of the African region and improvement of the quality of the life of the people, through rigorous education/training and outreach programmes thereby empowering and informing the people on Space for Development."[23]

---

[17]www.the-star.co.ke/news/2018/05/07/kenyas-first-satellite-set-for-launch-into-space-on-friday_c1755040, July 3, 2018.

[18]iHUB started in 2010 as an initiative that "aims to foster innovation and entrepreneurship within the Kenyan community, with a focus on Web and mobile services", see: https://ihub.co.ke/.

[19]m:Lab East Africa started in 2010 and provides a platform for mobile entrepreneurship, business incubation, developer-training and application-testing. One of the partner institutions is the University of Nairobi, webfoundation.org/projects/mlab-east-africa.

[20]@iLabAfrica is a centre in ICT innovation and development based at Strathmore University, a private university in Nairobi. It was established to address the Millennium Development Goals (MDGs) and to contribute toward Kenya's Vision 2030; see: ilabafrica.ac.ke/.

[21]www.konzacity.go.ke/the-vision/history/.

[22]UNESCO report, page 70.

[23]Agbaje, Ganiyu I.; United National Regional Centre for Space Science and Technology Education in Africa: Achievements, Opportunities, Challenges, and the Future, Environment and Ecology Research 5 (5): 386-394, 2017, here: 389.

ARCSSTE-E runs several educational programs such as a post-graduate diploma program and a Masters program in Space Science and Technology as well as the Space Education Outreach Programme (SEOP), which is aimed at stimulating the interest of students and teachers in primary, secondary, tertiary institutions and the general public in space science and technology. There are Space Education Workshops, the World Space Week Celebration and Workshops on Robotics. ARCSSTE-E has founded over 100 space clubs in schools and has developed curricula for space science education in primary and secondary schools.[24]

The National Space Research and Development Agency (NASDRA) was established in 1999 and the first Nigerian earth observation satellite (NigeriaSat-1) was developed in 2003 followed by three other satellites—NigeriaSat-2, NigeriaSat-X and NigComSat-1R—until 2011.[25] The National Space Research and Development Act (2010) was basis of the NASDRA strategy to establish six centers: the Centre for Basic Space Science and Astronomy, the National Centre for Remote Sensing, the Centre for Satellite Technology Development, the Centre for Geodesy and Geodynamics, the African Regional Centre for Space and Technology Education and the Centre for Space Transport and Propulsion.[26] The benefits of earth observation through usage of satellites is of significant importance for Nigeria as it affects surveillance and security, agriculture, disaster management and national planning purposes like population control and urban planning.[27]

In 2009, Nigeria launched the Vision 2020: Economic Transformation Blueprint with the clear objective that "by 2020 Nigeria will be one of the 20 largest economies in the world able to consolidate its leadership role in Africa and establish itself as a significant player in the global economic and political area."[28] In 2011, the Federal Ministry of Science and Technology set up the National Science, Technology and Innovation Fund and in 2014 the National Research and Innovation Council was set up.

The emphasis in STI is on space science and technology, biotechnology and renewable energy technologies.[29] One of the key goals of the Science, Technology and Innovation Policy is to develop an endogenous capability in launching and exploiting Nigeria's own satellites for telecommunications and research.[30] Furthermore, the Nigerian government has announced an ambitious space program including plans to launch an astronaut into space by 2030.[31]

---

[24]ibid: 390-392.

[25]http://nasrda.gov.ng/en/.

[26]Etomi, George and partners: Overview of space exploration in Nigeria: challenges and solutions for the effective operation of the sector, www.lexology.com/library/detail.aspx?g=f3a24bfa-5269-4cc3-9de2-cadb36a29018 March 22, 2018.

[27]ibid.

[28]www.nv2020.org/.

[29]UNESCO report, page 493.

[30]ibid.

[31]Monks, Kieron: Nigeria plans to send an astronaut to space by 2030, https://edition.cnn.com/2016/04/06/africa/nigeria-nasrda-space-astronaut/index.html, April 6, 2016.

1.6 South Africa

South Africa was one of the first countries, which were active in space exploration, starting with the South African Astronomical Observatory in 1820. Satellites were tracked primarily for weather and climate research and land use mapping. The first space program was launched in the 1980s, the first satellite, Sunsat-1, in 1999. The South African National Space Agency (SANSA) was established in December 2010 by the National Space Agency Act. In 2012 the vision of the National Development Plan is for South Africa to become a diversified economy firmly grounded in STI by 2030. This transition is guided by the Ten-Year Innovation Plan 2008–2018 and its five key topics: biotechnology and the bio-economy; space; energy security; global change; and understanding of social dynamics.

In 2012, South Africa also decided to host a 1.5 billion Euro project to build the world's largest radio telescope in South Africa and Australia: The African Very Long Baseline Interferometry Network (AVN) and the Square Kilometre Array (SKA)—this is bringing significant opportunities for research collaboration, attracting leading astronomers and researchers at all stages of their careers to work in Africa; it is worth noting that South African astronomers co-authored 89% of their publications with foreign collaborators during 2008–2014.[32] Eleven countries are currently members of the SKA Organization—Australia, Canada, China, Germany, India (associate member), Italy, New Zealand, South Africa, Sweden, the Netherlands and the United Kingdom.

Through the Department of Science and Technology (DST), South Africa has entered into 21 formal bilateral agreements with other African countries in science and technology since 1997, with a focus on Space Science with Botswana (2005), Egypt (1997), Ghana (2012) Kenya (2004), Mozambique (2006), Namibia (2005), Nigeria (2001), Rwanda (2009), Uganda (2009) and Zambia (2007).[33] Botswana, Ghana, Kenya, Mozambique, Namibia and Zambia are also partners in the AVN and SKA-project. Since 2005, the African SKA Human Capital Development Programme has awarded about 1000 grants for studies in astronomy and engineering from undergraduate to post-doctoral level, while also investing in training programs for technicians. Astronomy courses are being taught because of the SKA Africa project in Kenya, Mozambique, Madagascar and Mauritius, which has had a radio telescope for many years, and are soon to start in other countries.[34]

A great success was the launch of the DST "Technology Top 100" internship program in 2012, which places unemployed science, technology and engineering graduates in high-tech companies; in 2013 and 2014, one in four of the 105 interns were offered permanent employment with their host companies at the end of the one-year program; in 2015, a further 65 candidates were placed with companies in the Gauteng and Western Cape Provinces; it is planned to expand the network of

---

[32]UNESCO report, page 555.
[33]UNESCO report, page 556.
[34]http://www.ska.ac.za (July 1, 2018).

private firms involved in the program.[35] This will help to "promote sustained, inclusive and sustainable economic growth, full and productive employment and decent work for all" (SDG 8).

1.7 Zimbabwe

The Second Science and Technology Policy, launched in June 2012, cites sectorial policies with a focus on biotechnology, ICTs, space sciences, nanotechnology, indigenous knowledge systems, technologies yet to emerge and scientific solutions to emergent environmental challenges.[36]

In July 2018, the Zimbabwe National Geospatial and Space Agency (ZINGSA) was launched. The Agency is headed by the Ministry of Higher Education, Science and Technology Development and is structured in five departments: Space Operations and Launch Services, Space Science, Space Engineering, Geospatial and Earth Observation and Finance and Administration. ZINGSA will focus on four priority areas: Renewable Energy Mapping, Geological Minerals Mapping, Wildlife Surveillance and Health.[37]

2 Current Developments and Outlook

Some countries on the African continent are already taking steps to conduct space research and to train young people in science, technology, engineering and mathematics. The conditions in each of these countries to pursue their goals are extremely difficult. It therefore appears likely to seek solutions and join forces in an alliance of states: selected transnational initiatives are now summarized in a short overview.

Science, Technology and Innovation Strategy for Africa: STISA-2024.

In June 2014, the 23rd Ordinary Session of African Union Heads of State and Government Summit agreed to a 10-year Science, Technology and Innovation Strategy for Africa (STISA-2024). STISA-2024 has been developed during an important period when the African Union was formulating a broader and long-term AU Agenda 2063. STISA-2024 is the first of the ten-year incremental phasing strategies to respond to the demand for science, technology and innovation to impact across critical sectors such as agriculture, energy, environment, health, infrastructure development, mining, security and water. The strategy lists six priority areas that contribute to the achievement of the AU Vision: "An integrated, prosperous and peaceful Africa driven and managed by its own citizens and

---

[35]ibid.
[36]UNESCO report, page 562.
[37]Space in Africa: Zimbabwe Space Agency is launching tomorrow – all you need to know; https://africanews.space/zimbabwe-space-agency-is-launching-tomorrow-all-you-need-to-know/, July 9, 2018.

representing a dynamic force in the international arena".[38] These priority areas are: Eradication of Hunger and Achieving Food Security; Prevention and Control of Diseases; Communication (Physical and Intellectual Mobility); Protection of our Space; Live Together- Build the Society; and Wealth Creation.[39] "Space presents a unique opportunity for the continent to collectively address socio-economic development issues through derived services such as Earth Observation, Navigation and Positioning, Satellite Communication Space Science and Astronomy."[40]

2.1 African Space Policy

During the Second Ordinary Session for the Specialized Technical Committee Meeting on Education, Science and Technology (STC-EST) in October 2017 the African Space Policy: Towards Social, Political and Economic Integration was adopted.[41] Like in STISA-24, the AU Agenda 2063 is used as a framework with "the following key drivers: promoting science, technology and innovation; investing in human capital development; managing natural resources in a sustainable manner; effective private and public sector development and the promotion of public-private partnerships; innovative resource mobilization."[42]

The committee pointed out, that space science and technology is to be seen as basis for help on the way to a knowledge-based economy: "Space exploration presents a unique opportunity for cooperation in using and sharing enabling infrastructure and data towards the proactive management of disease, natural resources and the environment, responses to natural hazards and disasters, weather forecasting, climate change mitigation and adaptation, agriculture and food security, peace-keeping missions and conflict resolution."[43] Due to the high costs of space exploration and the long duration to build corresponding infrastructure, the committee is aware of the contribution of existing national programs towards a continental program. The challenge is to bring all these regional initiatives together "to create synergized, complementary programmes to foster collective actions towards Africa's development, and eventually enable the continent to be a global space player."[44]

From this two policy goals follow:

1. "To create a well-coordinated and integrated African space programme that is responsive to the social, economic, political and environmental needs of the continent, as well as being globally competitive.

---

[38]Science, Technology and Innovation Strategy for Africa, page 11; https://au.int/sites/default/files/newsevents/workingdocuments/33178-wd-stisa-english_-_final.pdf.
[39]ibid, page 10.
[40]ibid, page 23.
[41]African Union: African Space Policy: Towards Social, Political and Economic Integration: https://au.int/sites/default/files/newsevents/workingdocuments/33178-wd-african_space_policy_-_st20444_e_original.pdf, June 16, 2018.
[42]ibid, page 3.
[43]ibid, page 5.
[44]ibid, page 6.

2. To develop a regulatory framework that supports an African space programmes and ensures that Africa is a responsible and peaceful user of outer space."[45]

With the following objectives: addressing user needs; accessing space services; developing the regional market; adopting good governance and management; coordinating the African space arena; promoting intra-Africa and other international cooperation.

The following two initiatives are aimed in exactly this direction.

African Association of Remote Sensing of the Environment

The African Association of Remote Sensing of the Environment (AARSE) was founded in 1992. The AARSE international conference is conducted biennially across Africa and is the largest forum on the continent for researchers on remote sensing technologies and geospatial information science. In 1996, the first AARSE conference was held in Zimbabwe. Moreover, AARSE has also been organizing short courses from 3 days to 2 weeks in various parts of Africa. The main objective of AARSE conferences is to bring together scholars and professionals from the African and international community to present latest achievements, discuss challenges and share experiences. Conference programs usually feature keynote speeches delivered by leading scholars, technical sessions with reports of the latest research outcomes, discussion sessions on operational topics such as capacity building, Spatial Data Infrastructure (SDI) or Space Policy.[46]

African Initiative for Planetary and Space Sciences

In 2016, on the occasion of the 35th International Geological Congress in Cape Town, South Africa, an African Initiative for Planetary and Space Sciences[47] was proposed to foster Space Science in Africa. Besides a bilingual web presence, where the group of researchers tries to collect the widespread data of initiatives, calls, scientific symposia etc., they promote an initial investment for a program that prioritizes M.Sc. and Ph.D. scholarships, temporary study-abroad fellowships for young researchers and guest visits of junior and senior researchers to Africa for knowledge transfer.[48] They also point out that "African institutions must be made more aware of the resources readily available to them. These include data released by NASA and European Space Agency planetary missions along with tools and tutorials on how to process the data. The rise of open access journals offers African researchers greater access to the scientific literature, which will help future researchers on the continent."[49]

---

[45]ibid, page 8.
[46]www.africanremotesensing.org, July 2, 2018.
[47]https://africapss.org, July 2, 2018.
[48]"Africa Initiative for Planetary and Space Science" by David Baratoux and others, June 14, 2017 https://eos.org/opinions/africa-initiative-for-planetary-and-space-sciences; June 6, 2018.
[49]ibid.

3 Conclusion

This paper summarizes the evidence on aerospace research and related measures that can support African countries to reach 4 of 17 SDGs. Especially SDG 17 corresponds with one of the objectives of the African Space Policy: To promote the African-led space agenda through mutually beneficial partnerships that will include the aspects: intra-continental partnerships, international partnerships, partnerships across all sectors, equitable partnerships, ensure a reasonable and significant financial and/or social return and to influence international agreements.[50] SDG 17 includes the aspects "to enhance North-South, South-South and triangular regional and international cooperation on and access to science, technology and innovation and enhance knowledge sharing on mutually agreed terms, including through improved coordination among existing mechanisms, in particular at the United Nations level, and through a global technology facilitation mechanism. Promote the development, transfer, dissemination and diffusion of environmentally sound technologies to developing countries on favourable terms, including on concessional and preferential terms, as mutually agreed. Fully operationalize the technology bank and science, technology and innovation capacity-building mechanism for least developed countries by 2017 and enhance the use of enabling technology, in particular information and communications technology."[51] Number 70 of the objectives of the 2030 Agenda names one of the mechanisms leading in the same direction: "We hereby launch a Technology Facilitation Mechanism which was established by the Addis Ababa Action Agenda in order to support the Sustainable Development Goals. The Technology Facilitation Mechanism will be based on a multi-stakeholder collaboration between Member States, civil society, the private sector, the scientific community, United Nations entities and other stakeholders and will be composed of a United Nations inter-agency task team on science, technology and innovation for the Sustainable Development Goals, a collaborative multi-stakeholder forum on science, technology and innovation for the Sustainable Development Goals and an online platform."[52]

We recognize that some African countries have set up their own national space programs and have taken impressive measures. The desire among these countries for cooperation is tangible. In addition to this, supporting cooperation between these fragmented initiatives under the umbrella of the African Union should be possible. South-South as well as North-South partnerships are crucial for successful programs as space exploration is complex and too expensive. Only in conjunction with several countries, significant progress can be made. It is the aim of this paper to show that the great challenges of our time can only be solved through joint efforts. Therefore I believe in the well-chosen topic of 2018 World's Space Week, held on October 4–10: "Space Unites The World".[53]

---

[50]African Space Policy, page 14.
[51]Transforming our world: the 2030 Agenda for Sustainable Development, page 25.
[52]ibid, page 30.
[53]http://www.worldspaceweek.org/nations/..

Author Biography

Christine Müller studied East European History, Political Science and German Studies in Tübingen, Heidelberg, Warsaw and Krakow. She is working in science management since 2006 at the Universities of Heidelberg, Bonn, Stuttgart and Siegen. From 2011 until 2014 she was lecturer with specialized functions at the German Academic Exchange Service (DAAD) office in Warsaw, where she was responsible for organizing scientific networking events within the framework of Poland's accession to the European Space Agency. In 2016, Christine Müller was one of the speakers at a symposium about the "Current Situation and Development of Further Education and Research in Vocational Education and Training in Sub-Saharan Africa" at the Namibia's University of Science and Technology (NUST) in collaboration with University of Rostock. From 2016 until 2018, she was also in charge for the University of Siegen's partnership with Dedan Kimathi University of Science and Technology (DeKUT) in Nyeri, Kenia. Since April 2018 Christine Müller is Head of the Department of "European and International Networks" in the International Office at the University of Bonn.

Towards the Sustainable Development Goals in Africa: The African Space-Education Ecosystem for Sustainability and the Role of Educational Technologies

André Siebrits⊙ and Valentino van de Heyde⊙

Abstract

The United Nations Sustainable Development Goals (SDGs) encapsulate the collective desire of the world to eliminate the worst miseries of poverty and to set itself on a sustainable growth path. This chapter considers the origin of sustainable development in global discourse, and unpacks its education-related goals and targets. The specific focus is placed on tertiary education and it is argued that the SDGs depend on both regional and national strategies to succeed. Related questions of educational quality, and the role of Information and Communications Technology (ICT), are considered. The progress of the education-related SDGs is investigated, especially in relation to Internet and mobile phone penetration in Africa, and tertiary enrolment rates. ICTs, and specifically e-learning, are discussed as a means of helping the challenges related to the massification of the tertiary education sector in Africa. The way in which space supports education is considered as only one pillar of what in reality is an interrelated symbiotic relationship, which feeds back into producing space-related skills to advance the African space sector. Recommendations in light of space, education, and sustainable e-learning are discussed, and the importance of recognising the value of space in achieving the SDGs is emphasised.

A. Siebrits (✉)
Department of Political Studies, University of Cape Town,
Rondebosch, South Africa
e-mail: SBRAND003@myuct.ac.za

V. van de Heyde
Department of Physics and Astronomy, University of the Western Cape,
Bellville, South Africa
e-mail: vvandeheyde@uwc.ac.za

© Springer Nature Switzerland AG 2019
A. Froehlich (ed.), *Embedding Space in African Society*, Southern Space Studies,
https://doi.org/10.1007/978-3-030-06040-4_10

Science knows no country, because knowledge belongs to humanity, and is the torch which illuminates the world. Science is the highest personification of the nation because that nation will remain the first which carries the furthest the works of thought and intelligence.

(Louis Pasteur)[1]

1   Introduction

The concept of sustainable development (SD) first came to international prominence at the 1972 Stockholm United Nations (UN) Conference on the Human Environment.[2] The Stockholm Conference was the first major UN gathering focusing on international environmental issues, and as such it "marked a turning point in the development of international environmental politics".[3] While SD had not yet entered into global discourse, it was noted in the conference report that "[t]he problem was how to reconcile those legitimate immediate requirements [of developing countries] with the interests of generations yet unborn".[4] This was the turning point in the sense that it was recognised that the issues of development and the environment—previously dealt with separately—could be managed more beneficially jointly, since they were in reality inseparable.

The "standard definition" of SD originated 15 years later through the World Commission on Environment and Development, launched by the UN in 1982, and chaired by former Norwegian Prime Minister Gro Harlem Bruntland.[5] The Bruntland Commission drew on both the 1972 Stockholm Conference and the 1980 World Conservation Strategy of the International Union for the Conservation of Nature, which promoted "conservation as a means to assist development and specifically for the sustainable development and utilization of species, ecosystems, and resources".[6] In the resulting 1987 Bruntland Commission report, entitled *Our*

---

[1]Quoted in Carl C. Gaither and Alma E. Cavazos-Gaither, *Scientifically Speaking: A Dictionary of Quotations* (Bristol: IOP Publishing, 2000), 199.
[2]Sustainable Development Commission, *History of SD*, http://www.sd-commission.org.uk/pages/history_sd.html (accessed August 4, 2018).
[3]United Nations Department of Economic and Social Affairs, *United Nations Conference on the Human Environment* (*Stockholm Conference*), https://sustainabledevelopment.un.org/milestones/humanenvironment (accessed August 4, 2018).
[4]United Nations, *Report of the United Nations Conference on the Human Environment* (New York: United Nations, 1972), 45. http://www.un.org/ga/search/view_doc.asp?symbol=A/CONF.48/14/REV.1 (accessed 4 August, 2018).
[5]Robert W. Kates, Thomas M. Parris, and Anthony A. Leiserowitz, "What Is Sustainable Development? Goals, Indicators, Values, and Practice," *Environment: Science and Policy for Sustainable Development* 47, no. 1 (2005): 1–13.
[6]Ibid., 2.

Common Future, SD was accordingly conceptualised as "meet[ing] the needs of the present without compromising the ability of future generations to meet their own needs".[7] The report went on to argue that

> The environment does not exist as a sphere separate from human actions, ambitions, and needs, and attempts to defend it in isolation from human concerns have given the very word 'environment' a connotation of naivety in some political circles. The word 'development' has also been narrowed by some into a very limited focus, along the lines of 'what poor nations should do to become richer', and thus again is automatically dismissed by many in the international arena as being a concern of specialists, of those involved in questions of 'development assistance'. But the 'environment' is where we all live; and 'development' is what we all do in attempting to improve our lot within that abode. The two are inseparable.[8]

From here, the concept of SD "formed the basis" of the 1992 UN Conference on Environment and Development (UNCED) in Rio de Janeiro (known as the 'Earth Summit').[9] There it was placed at the centre of the subsequent Rio Declaration, which reaffirmed that "[i]n order to achieve sustainable development, environmental protection shall constitute an integral part of the development process and cannot be considered in isolation from it".[10] It was at Rio where SD became recognised as a major global challenge, with this recognition being further reaffirmed at the 2002 Johannesburg World Summit on Sustainable Development.[11]

It was thus in many ways a natural evolution to follow up the 2000–2015 UN Millennium Development Goals (MDGs) with the 2015–2030 UN Sustainable Development Goals (SDGs). Originating from the 2000 Millennium Summit, the eight MDGs constituted "a type of global report card" around which global partnerships could rally to address the most pressing social priorities.[12] These eight goals were focused on eradicating extreme poverty and hunger, achieving universal primary education, promoting gender equality and empowerment of women, reducing child mortality, improving maternal health, combating diseases like malaria and HIV, ensuring environmental sustainability, and fostering a global partnership for development.[13] While not all of these goals or their individual objectives were achieved—a "serious, regrettable, and deeply painful" shortfall— notable and worthy progress was made.[14]

---

[7]Brundtland Commission, *Report of the World Commission on Environment and Development: Our Common Future* (New York: United Nations, 1987), 30. http://www.un-documents.net/our-common-future.pdf (accessed 5 August, 2018).

[8]Ibid., 7.

[9]Sustainable Development Commission, *History of SD.*

[10]The United Nations Conference on Environment and Development, *The Rio Declaration on Environment and Development* (New York: United Nations, 1992), 2. http://www.unesco.org/education/pdf/RIO_E.PDF (accessed 6 August, 2018).

[11]Sustainable Development Commission, *History of SD.*

[12]Jeffrey D. Sachs, "From Millennium Development Goals to Sustainable Development Goals," *Lancet*, 2012, no 379: 2206.

[13]United Nations, *Background*, http://www.un.org/millenniumgoals/bkgd.shtml (accessed August 5, 2018).

[14]Sachs, "From Millennium Development Goals to Sustainable Development Goals," 2206.

For this reason, the consensus emerged that a new set of goals was required for the post-2015 development agenda, which would build on the MDGs while moving from having SD as an individual goal to placing it at the centre of an expanded set of 17 goals and 169 associated targets.[15] These SDGs are predicated on people, planet, prosperity, peace, and partnership. The SDGs are a marked improvement on the MDGs in the sense that they "aim to cover the whole sustainable development universe, which includes basically all areas of the human enterprise on Earth" and, in contrast to the MDGs in which "'silo' goals encouraged silo policies and did not make links and trade-offs across areas explicit", the SDGs are more connected, reflecting "the recognition by the international community of the importance of links among the goals".[16] Having applied network analysis techniques to create a map of the SDG network, illustrating the connections between the various goals and indicators, Le Blanc identified five of the 16 goals (not counting SDG 17, which deals with implementation) as constituting the 'core' of the interlinked SDG network.[17] These are the SDGs with the most connections with other goals: 1 (end poverty in all its forms everywhere), 2 (end hunger, achieve food security and improved nutrition and promote sustainable agriculture), 4 (ensure inclusive and equitable quality education and promote lifelong learning opportunities for all), 10 (reduce inequality within and among countries), and 12 (ensure sustainable consumption and production patterns).[18]

In this chapter, we will focus on SDG 4, since it is clear that education is one of the areas with the most potential to impact the world's ability to achieve the desired development by 2030. As former South African president Nelson Mandela succinctly put it, "[e]ducation is the most powerful weapon we can use to change the world".[19] This is further borne out by Le Blanc's identification of the strengths of the connections between the various SDGs, and of all links, the strongest were between education and gender (goal 5), and between poverty and inequality.[20] In total, education was linked to seven other SDGs through their various targets.[21] From this, it is clear that individual SDGs not only include their own 'core' targets, but also what Le Blanc terms "extended" targets which are related but fall under other SDGs.[22] For this reason, it is important to include these extended targets in any analysis of the SDGs. Our analysis will, however, only be concerned with

---

[15]United Nations Department of Economic and Social Affairs, *Transforming our world: the 2030 Agenda for Sustainable Development*, https://sustainabledevelopment.un.org/post2015/transformingourworld (accessed August 6, 2018).

[16]David Le Blanc, "Towards integration at last? The sustainable development goals as a network of targets," *DESA Working Paper*, 2015, no. 141: 11–15.

[17]Ibid., 3.

[18]Ibid.

[19]Nelson Mandela Foundation, *Address by Nelson Mandela at launch of Mindset Network*, July 16, 2003, http://www.mandela.gov.za/mandela_speeches/2003/030716_mindset.htm (accessed August 5, 2018).

[20]Le Blanc, "Towards integration at last?," 3.

[21]Ibid., 6.

[22]Ibid., 9.

tertiary education, for reasons we will highlight later, and within tertiary education we will concentrate on learning technologies.

Our second focus—Africa—is motivated by the dire needs which remain in the context of the SDGs. While the African continent made great improvements in terms of the MDGs, it nevertheless remains true that "progress has been very slow".[23] It is therefore encouraging that, alongside the central focus on SD, the post-2015 development agenda "strongly placed" the focus on Africa.[24] It is also encouraging that Africa's own Agenda 2063, which sets out the "strategic framework for the socio-economic transformation of the continent",[25] expands on and domesticates issues of sustainable development. Education has been identified as one of four intersecting issues between the SDGs and Agenda 2063, making it a pillar for Africa's future development (the other intersecting issues are technology, women, and inclusiveness).[26] We will relate our discussion to, and use examples from, the entire African continent since we argue that the SD and educational challenges confronting Africa cannot be resolved in isolation.

Our third focus is on the role of space (and its related activities, technologies, and data) in helping to achieve SDG 4 in Africa. The UN Office for Outer Space Affairs (UNOOSA) identifies two key roles of space in terms of education, namely facilitating educational programme delivery, and motivating and encouraging students to pursue the sciences.[27] In relation to the former, the role of space in supporting various technologies is emphasised, such as the provision of Internet services, and tele-education using video conferencing, web-based learning materials, and interactive space simulations. In relation to the second key role of space in terms of education, space itself acts as a motivator to spark curiosity and interest in the sciences, and can act as a vehicle for mobilising future generations of scientists. This is especially critical for Africa since, as Pasteur's comment illustrated, scientific skills are the foundation of national prosperity. In relation to the role of space in supporting SDG 4, we will make use of our Symbiotic Education Enabler Model for Space (SEEMS) to highlight that while space can advance this SDG both through supporting programme delivery, and through motivating students to pursue sciences (and of course space studies), these in turn can also advance the African space sector by developing both the cadre of future professionals that will comprise this sector as well as urgently needed skills in science and education provision. In this way, we highlight the full complexity and symbiotic nature of education and educational technologies, Africa, and space.

---

[23]Godwell Nhamo, "New Global Sustainable Development Agenda: A Focus on Africa," *Sustainable Development* 25, no. 3 (2017): 7.

[24]Ibid., 8.

[25]African Union Commission, *What is Agenda 2063?*, https://au.int/en/agenda2063 (accessed August 5, 2018).

[26]Nhamo, "New Global Sustainable Development Agenda," 9.

[27]United Nations Office for Outer Space Affairs, *Benefits of Space: Education*, 2018, http://www.unoosa.org/oosa/en/benefits-of-space/education.html (accessed August 5, 2018).

This chapter will accordingly be structured as follows. First, SDG 4 will be analysed, including the core targets and indicators, as well as the extended targets and indictors that relate to our focus on higher education. The particular goals and targets of Agenda 2063 will also be considered, and as an example, so will the targets related to higher education in South Africa's own National Development Plan 2030. Following this, ways in which this goal is already implemented in general in Africa at large will be explored, including the progress of the implementation of SDG 4. As part of narrowing our focus, we will pay particular attention to the role of e-learning in tertiary education for SD in Africa. Then, the role of space in this existing implementation will be considered, followed by a discussion of the future advancement of SDG 4 via space, including the symbiotic interrelationship mentioned above.

2 Sustainable Development and Education in Africa

Education can be the difference between a life of grinding poverty and the potential for a full and secure one; between a child dying from preventable disease, and families raised in healthy environments; between orphans growing up in isolation, and the community having the means to protect them; between countries ripped apart by poverty and conflict, and access to secure and sustainable development.

(Nelson Mandela and Graça Machel)[28]

Education has been tied to SD since the beginning. Principle 19 of the Declaration of the United Nations Conference on the Human Environment (stemming from the 1972 Stockholm Conference), states that

Education in environmental matters, for the younger generation as well as adults, giving due consideration to the underprivileged, is essential in order to broaden the basis for an enlightened opinion and responsible conduct by individuals, enterprises and communities in protecting and improving the environment in its full human dimension. It is also essential that mass media of communications avoid contributing to the deterioration of the environment, but, on the contrary, disseminate information of an educational nature on the need to protect and improve the environment in order to enable man to develop in every respect.[29]

The role of education as the foundation for "enlightened opinion and responsible conduct" is critical to the SDGs as well. Without this, it is dubious whether any "sustainable trajectory"[30] can endure. As the Bruntland Commission report put it, "[t]he changes in human attitudes that we call for depend on a vast campaign of education, debate, and public participation".[31] For this reason, the need for

---

[28]Quoted in Clayton R. Wright, Gajaraj Dhanarajan, and Sunday A. Reju, "Recurring Issues Encountered by Distance Educators in Developing and Emerging Nations," *International Review of Research in Open and Distance Learning* 10, no. 1 (2009): 1–2.

[29]United Nations, *Report of the United Nations Conference on the Human Environment*, 5.

[30]Sachs, "From Millennium Development Goals to Sustainable Development Goals," 2206.

[31]Brundtland Commission, *Report of the World Commission on Environment and Development*.

including basic education in the MDGs was clear, and thus goal 2 called for achieving universal primary education, by ensuring that it becomes possible for all children everywhere to complete a full course of primary schooling.[32] While meaningful progress was achieved in this regard (91% enrolment in primary education in developing countries in 2015, up from 83% in 2000), a large gap persisted whereby 57 million children of primary school age remained out of school in 2015. Without achieving this most basic level of education, the prospects for lifting communities out of poverty remain dubious, as is achieving SD. Therefore, understandably so, primary education remains a cornerstone of SDG 4.

In contrast to the MDGs, the SDGs expanded the focus on education beyond the primary level to include early childhood, secondary, tertiary, and vocational education, as well as learning environments and infrastructure—a very positive development. As the United Nations Development Programme posits, "[t]his goal ensures that all girls and boys complete free primary and secondary schooling by 2030" and "also aims to provide equal access to affordable vocational training, to eliminate gender and wealth disparities, and achieve universal access to a quality higher education".[33] This full-spectrum approach thus takes a more holistic view on education.

The expansion of SDGs from the MDGs was not, however, met without criticism, and while some were more generous by calling the development of a much broader range of goals "an ambitious challenge",[34] others were more pointed in their criticism. *The Economist* published an article entitled *The 169 Commandments*, in which they argued that the "proposed sustainable development goals would be worse than useless", calling the SDGs a "mess", "unfeasibly expensive", "narrow", a "distraction", and "ambitions on a Biblical scale, and not in a good way".[35] Others echoed this sentiment by averring that the "SDGs are far too numerous and many issues emerge from them", including challenges related to communicating them to the public, monitoring, verifying, and reporting on progress, availability of data, individual and institutional capacity, and "reporting fatigue".[36] Despite this, the fact that the SDGs are better integrated with stronger links between individual goals and targets than the MDGs helps to illustrate that achieving SD requires a broad-based approach, and a "cross-cutting means of implementation", provided for in SDG 17.[37]

---

[32]United Nations, *Goal 2: Achieve Universal Primary Education*, http://www.un.org/millenniumgoals/education.shtml (accessed August 5, 2018).
[33]United Nations Development Program, *Goal 4: Quality Education*, 2018, http://www.za.undp.org/content/south_africa/en/home/sustainable-development-goals/goal-4-quality-education.html (accessed August 5, 2018).
[34]Le Blanc, "Towards integration at last?," 1.
[35]The Economist, *The 169 commandments*, March 26, 2015, https://www.economist.com/leaders/2015/03/26/the-169-commandments (accessed August 5, 2018).
[36]Nhamo, "New Global Sustainable Development Agenda," 12.
[37]Le Blanc, "Towards integration at last?," 2.

Because primary education, and to a lesser extent secondary education, have received the bulk of attention in relation to the SDGs, our approach is aimed at the tertiary sector.[38] While sub-Saharan Africa has the lowest tertiary enrolment rate of any region on Earth,[39] African countries are already grappling with tremendous challenges in this sector, and will most likely face even more severe challenges in future. Historically, there has been a trend for most public institutions of higher education in Africa to enrol students "in excess of their capacity", and pressures stemming from "rapidly increasing secondary school graduates" will increase in future (since many of these will want to attend tertiary education), presenting "a compelling need to further increase tertiary enrolment".[40] Success on primary and secondary education levels, as supported by the MDGs in the case of the former, and the SDGs in the case of both, coupled with rapid population growth in many African countries, means that "the social and political pressures to admit large numbers of students are inescapable".[41] Such massification of tertiary education is undoubtedly a positive development on a national scale, given the undeniable "importance of higher education in addressing development challenges" and addressing issues of equity in relation to access to education, but on an institutional level such massification can bring about serious challenges related to institutional carrying capacity.[42] Coupled with this, many African countries report high attrition rates, especially in relation to mathematics (as high as 95% in the Central African Republic).[43] Neither the space sector nor SD will benefit if this continues to remain the case, and solutions are urgently needed if tertiary education is to fulfil its developmental, and indeed emancipatory, potentials. For this reason we have placed our focus squarely on this sector, given the inevitable challenges related to massification that will arise as more secondary school graduates clamour for tertiary education, as evidenced by the recent *Fees Must Fall* student protest in South Africa, which was an "uprising against lack of access to, and financial exclusion from, higher education".[44]

Accordingly, all core and extended targets and indicators relating to tertiary education (as pointed out by Le Blanc) are listed in Table 1 (plus our addition of SDG 17), with those relating to our discussion identified in bold. Excluded from

---

[38]The reason for the use of the term 'tertiary education' is because, as Mohamedbhai notes, "The terms 'higher education' and 'tertiary education' are often used interchangeably to denote post-secondary education. 'Tertiary education' is more encompassing and covers all post-secondary education. 'Higher education' normally refers to education leading to a degree. Most international statistics are in terms of tertiary education." Goolam Mohamedbhai, "Massification in Higher Education Institutions in Africa: Causes, Consequences, and Responses," *International Journal of African Higher Education* 1, no. 1 (2014): 62.

[39]Ibid., 59.

[40]Ibid.

[41]Ibid., 66.

[42]Ibid., 61–67.

[43]Ibid., 71.

[44]Susan Booysen, ed., *Fees Must Fall: Student revolt, Decolonisation and Governance in South Africa* (Johannesburg: WITS University Press, 2016). Quote from Project Muse, Fees Must Fall: Student revolt, Decolonisation and Governance in South Africa, *Summary*, 2018, https://muse.jhu.edu/book/50547 (accessed August 11, 2018).

Table 1 Sustainable development goals relating to education[a]

Targets	Indicators
Goal 4. Ensure inclusive and equitable quality education and promote lifelong learning opportunities for all	
3. By 2030, ensure equal access for all women and men to affordable and quality **technical, vocational and tertiary education, including university**	3.1. Participation rate of youth and adults **in formal and non-formal education and training** in the previous 12 months, by sex
4. By 2030, substantially increase the number of youth and adults who have **relevant skills, including technical and vocational skills**, for employment, decent jobs and entrepreneurship	4.1. Proportion of youth and adults with **information and communications technology (ICT) skills**, by type of skill
5. By 2030, eliminate gender disparities in education and ensure **equal access to all levels of education and vocational training** for the vulnerable, including persons with disabilities, indigenous peoples and children in vulnerable situations	5.1. Parity indices (female/male, rural/urban, bottom/top wealth quintile and others such as disability status, indigenous peoples and conflict-affected, as data become available) for all education indicators on this list that can be disaggregated
6. By 2030, ensure that all youth and a substantial proportion of adults, both men and women, achieve **literacy and numeracy**	6.1. Percentage of population in a given age group achieving at least a fixed level of proficiency in functional (a) literacy and (b) numeracy skills, by sex
7. By 2030, ensure that all learners acquire the **knowledge and skills needed to promote sustainable development**, including, among others, through **education for sustainable development** and sustainable lifestyles, human rights, gender equality, promotion of a culture of peace and non-violence, global citizenship and appreciation of cultural diversity and of culture's contribution to sustainable development	7.1. Extent to which (i) **global citizenship education** and (ii) **education for sustainable development**, including gender equality and human rights, are mainstreamed at all levels in: (a) national education policies, (b) curricula, (c) teacher education and (d) student assessment
A. **Build and upgrade education facilities** that are child, disability and gender sensitive and provide safe, non-violent, inclusive and **effective learning environments** for all	1. Proportion of schools with access to: (a) electricity; (b) the **Internet for pedagogical purposes**; (c) **computers for pedagogical purposes**; (d) **adapted infrastructure and materials for students with disabilities**; (e) basic drinking water; (f) single-sex basic sanitation facilities; and (g) basic handwashing facilities (as per the WASH indicator definitions)
C. By 2030, substantially increase the supply of **qualified teachers**, including through international cooperation for **teacher training** in developing countries, especially least developed countries and small island developing States	1. Proportion of teachers in: (a) pre-primary; (b) primary; (c) lower secondary; and (d) upper secondary education who have received at least the minimum organized teacher training (e.g. pedagogical training) pre-service or in-service required for teaching at the relevant level in a given country

(continued)

Table 1 (continued)

Targets	Indicators
Goal 3. Ensure healthy lives and promote well-being for all at all ages	
7. By 2030, ensure universal access to sexual and reproductive health-care services, including for family planning, **information and education**, and the integration of reproductive health into national strategies and programmes	7.1. Proportion of women of reproductive age (aged 15–49 years) who have their need for family planning satisfied with modern methods 7.2. Adolescent birth rate (aged 10–14 years; aged 15–19 years) per 1000 women in that age group
Goal 5. Achieve gender equality and empower all women and girls	
6. Ensure universal access to sexual and reproductive health and reproductive rights as agreed in accordance with the Programme of Action of the International Conference on Population and Development and the Beijing Platform for Action and the outcome documents of their review conferences	6.1. Proportion of women aged 15–49 years who make their own **informed decisions** regarding sexual relations, contraceptive use and reproductive health care 6.2. Number of countries with laws and regulations that guarantee women aged 15–49 years access to sexual and reproductive health care, information and **education**
Goal 13. Take urgent action to combat climate change and its impacts	
3. **Improve education**, awareness-raising and human and **institutional capacity** on climate change mitigation, adaptation, impact reduction and early warning	3.1. Number of countries that have **integrated mitigation, adaptation, impact reduction and early warning** into primary, secondary and **tertiary curricula** 3.2. Number of countries that have communicated the **strengthening of institutional, systemic and individual capacity-building** to implement adaptation, mitigation and technology transfer, and development actions
Goal 17. Strengthen the means of implementation and revitalise the global partnership for sustainable development	
6. Enhance North-South, South-South and triangular regional and international cooperation on and access to **science, technology and innovation** and enhance **knowledge sharing** on mutually agreed terms, including through improved coordination among existing mechanisms, in particular at the United Nations level, and through a global technology facilitation mechanism	6.1. Number of **science and/or technology** cooperation agreements and programmes between countries, by type of cooperation 6.2. Fixed **Internet broadband** subscriptions per 100 inhabitants, by speed
8. Fully operationalize the technology bank and **science, technology and innovation capacity-building** mechanism for least developed countries by 2017 and **enhance the use of enabling technology, in particular information and communications technology**	8.1. Proportion of individuals **using the Internet**

[a]United Nations Department of Economic and Social Affairs, *Sustainable Development Goals*, https://sustainabledevelopment.un.org/?menu=1300 (accessed August 5, 2018)

this list are the goals exclusively addressing primary education, scholarships, as well as extended goal 12.6, which relates to companies publishing sustainability reports and integrating sustainability information into their reporting cycle. The reason we include SDG 17 here is because of its targets relating to internet use and enabling technologies (particularly Information and Communications Technologies —ICTs). These play a critical role in facilitating education delivery and access across a continent with many remote communities which do not enjoy easy access to educational services often concentrated around urban areas.

Like the SDGs, Africa's Agenda 2063 sets out a range of aspirations for the continent, the first of which is a "prosperous Africa based on inclusive growth and sustainable development".[45] SD is thus incorporated and domesticated here as an integral foundation of prosperity. Coupled with this aspiration is the vision that by 2063, Africa will be a continent where "[w]ell educated and skilled citizens, underpinned by science, technology and innovation for a knowledge society is the norm and no child misses school due to poverty or any form of discrimination".[46] It is recognised that to achieve this, significant investments will be required so that

> Africa's human capital will be fully developed as its most precious resource, through sustained investments based on universal early childhood development and basic education, and sustained investments in higher education, science, technology, research and innovation, and the elimination of gender disparities at all levels of education. Access to post-graduate education will be expanded and strengthened to ensure world-class infrastructure for learning and research and support scientific reforms that underpin the transformation of the continent.[47]

In order to realise the 2063 aspirations, Agenda 2063 is broken down into a series of decade-long implementation plans. As such, the First Ten-Year Implementation Plan 2014–2023 sets out the priority areas, goals, and targets to be achieved by all African states and the continent as a whole. While the deadline for these is 2023, the longer-term nature of Agenda 2063 makes these goals and targets a good reflection of African views on domesticating the SDGs listed above. In relation to education, Table 2 sets out the targets for 2023. Some of the indicative strategies on national level to support these targets are included as well, since they echo the points above relating to the learning environments and ICTs. Once again, bold text indicates areas of particular concern in relation to our discussion.

In the final instance, as the national targets of Agenda 2063 suggest, neither the high-level SDGs nor mid-level continental Agenda 2063 can be successfully achieved without clear plans for national implementation. In this regard, South Africa can serve as an example, where the National Development Plan (NDP) 2030 sets out the "determined and measurable actions [required] from all social actors

---

[45]African Union Commission, *Agenda 2063: The Africa We Want—Popular Version* (Addis Ababa: African Union, 2016), 2. https://au.int/sites/default/files/documents/33126-doc-03_popular_version.pdf (accessed August 7, 2018).
[46]Ibid., 3.
[47]Ibid., 4.

Table 2 Agenda 2063 education-related goal, targets, and indicative strategies[a]

Aspiration 1. A prosperous Africa based on inclusive growth and sustainable development	Goal 2. Well educated citizens and skills revolution underpinned by science, technology and innovation	• **Priority area 1**. Education and STI driven skills revolution *National targets* (2023): 3. Increase number of **qualified teachers by at least 30% with focus on STEM** 5. At least 30% of secondary school leavers go into **tertiary education** with at least 40% being female 6. At least 70% of secondary school students not entering the tertiary sector are provided with a range of options for **further skills development** 7. At least 70% of the public perceive **quality improvements in education** at all levels *Continental Targets* (2023): 1. African Education Accreditation Agency is fully operational 2. Common continental education qualification system is in place 3. African e-University is established 4. Pan African University is consolidated with at least 25 satellite centres 5. African Education Observatory is fully operational 6. At least 50% of Member states have national accreditation systems in place by 2023 7. Framework for Harmonization of Teacher Education is completed by 2018

Some indicative strategies (*national*):
• Expand and improve **educational facilities/access** at the early childhood, Basic, Secondary, TVET and Tertiary levels with focus on **science, technology and innovation**
• Strengthen/establish network of **vocational training centres/incubators**
• Increase the **supply of qualified teachers/instructors** at all levels by improving **training capacity** and teacher/instructor incentives to ensure that they possess the relevant knowledge, skills and attitudes and motivation to teach effectively

(continued)

Table 2 (continued)

- Design/implement strategies to increase the level of incentives for teachers to ensure enhanced recruitment and retention of **qualified teachers**
- Expand/improve **educational infrastructure** at all levels to support **STEM/skills revolution** agenda
- Create a **conducing enabling environment** for the education sector that promotes/supports expansion in technical and analytical competencies, entrepreneurship and innovative skills of learners
- Provide public/community libraries to enhance learning, **access to information and knowledge**
- Strengthening the **learning infrastructure for STEM** that ensures increased incentives, access and quality learning for women and girls
- Develop/implement programmes to enhance the **capacity of science and technology institutions**
- Design/implement policies that provide options for further **education and skills training** for secondary school graduates who do not enter the tertiary sector
- Develop/implement programmes to govern tertiary institutions to **ensure quality education**
- Put in place policies to **nurture research and innovation culture**
- Develop/implement **ICT policies for educational institutions**

[a]African Union Commission, *Agenda 2063: The Future We Want—First Ten-Year Implementation Plan 2014–2023* (Addis Ababa: African Union, 2015), 51–53. https://au.int/sites/default/files/documents/33126-doc-ten_year_implementation_book.pdf (accessed August 7, 2018)

and partners across all sectors in society"[48] in order to realise the vision of a better life for all citizens, and acknowledges that "[a]chieving a competitive and sustainable economy will require a strong and effective system of innovation, science and technology".[49] Similarly, the NDP correctly recognises that "[h]igher education is the major driver of information and knowledge systems that contribute to economic development" and that it is "also important for good citizenship and for enriching and diversifying people's lives"—all cornerstones of SD.[50]

The NDP sets out national targets for all levels of education, and those relating to higher education are illustrated in Table 3. The NDP outlines a number of policy proposals to support these ambitious goals, but it is clear that these goals also rely to a great extent on primary and secondary levels of education.

This example of the NDP is important here for two reasons. First, it illustrates, in conjunction with the SDGs and Agenda 2063, that the SDGs are only one level in a three-tiered system encompassing global, regional, and national levels, all of which must be in alignment in order to realise SD. We argue that a critical facet of this is the extent to which these levels recognise the value of space in achieving their

---

[48]National Planning Commission, *The National Development Plan*, http://www.nationalplanningcommission.org.za/Pages/NDP.aspx (accessed August 11, 2018).

[49]National Planning Commission, *National Development Plan 2030: Our Future—Make It Work* (Pretoria: Presidency of South Africa, 2012), 326. http://www.nationalplanningcommission.org.za/Downloads/ndp-2030-our-future-make-it-work_0.pdf (accessed August 11, 2018).

[50]Ibid., 317.

Table 3 South African National Development Plan 2030 proposals for universities[a]

Universities
• Improve the qualifications of higher education academic staff. South Africa needs to increase the percentage of Ph.D. qualified staff in the higher education sector from the current 34% to over 75% by 2030
• Improve the **quality of teaching and learning**. University lecturers should be recognised teachers
• Increase the participation rate at universities by at least 70% by 2030 so that enrolments increase to about 1.62 million from 950,000 in 2010
• Increase the throughput rate for degree programmes to more than 75%. The number of graduates will increase from the combined total of 167,469 for private and public higher education institutions to 425,000 by 2030. As part of this target, the number of **science, technology, engineering and mathematics graduates** should increase significantly
• Increase the number of masters and Ph.D. students, including by supporting partnerships for research. By 2030 over 25% of university enrolments should be at postgraduate level
• Produce more than 100 doctoral graduates per million per year by 2030
• Double the number of graduate and postgraduate scientists and increase the number of African and women postgraduates, especially Ph.D.s, to improve research and innovation capacity and make university staff more representative
• Create a **learning and research environment that is welcoming** to all
• **Expand university infrastructure**. University enrolments have almost doubled since 1994 and **infrastructure has not kept up**. This has a major impact on the **quality of teaching and learning**. Student accommodation in universities needs urgent attention
• Develop uniform standards for **infrastructure and equipment** to support learning, promote equity and ensure that **learners doing similar programmes in different institutions receive a comparable education**
• Strengthen universities that have an embedded culture of research and development. They should be assisted to access private sector research grants (third stream funding) in addition to state subsidies and student fees, attract researchers, form partnerships with industry and **be equipped with the latest technologies**
• Provide performance-based grants to build capacity and develop centres or networks **of excellence** within and across institutions
• Offer **extra support** to underprepared learners to help them cope with the demands of higher education. Many individuals with poor schooling aspire to higher qualifications, but they are academically less prepared than their middle class counterparts. **Support programmes** should be offered and funded at all institutions
• Expand the use of **distance education**. The **advances in ICT** can help overcome the infrastructure limits to further expansion of higher education. Upfront investment is needed in **technology, curriculum design, quality assurance and monitoring**
• Private providers will continue to be important partners in the delivery of education and training at all levels. Ensuring the quality of private provision requires enabling regulation, quality assurance, and monitoring and evaluation of programmes

[a]Ibid., 319–320

targets and goals, something we will explore further in the section relating to role of space in implementing the educational SDGs. The second reason the example of the NDP is important is because it reflects a realisation that ICT has a critical role to play in achieving the ambitious goals set out for tertiary education. The areas where ICTs can support education are therefore indicated in bold text in Table 3. The UN Educational, Scientific, and Cultural Organisation (UNESCO) recognises both the importance of education for SD, and of ICTs for education, as encapsulated in the 2015 Incheon Declaration.

2.1 Tapping into ICTs for Expanding Education

The Incheon Declaration was issued at the 2015 World Education Forum held in South Korea, and sets out a new vision for education in its Education 2030 Framework for Action. This framework acknowledged that education systems face the enormous challenge of being both relevant and responsive to a range of factors, including changing labour markets, technological advances, poverty, widening inequality, political instability, environmental degradation, and many others.[51] Whilst grappling with these issues, education systems will have to accommodate "hundreds of millions" of new entrants by 2030.[52] Given that "[t]ertiary education and universities are critical for the education of future scientists, experts and leaders",[53] the issue of educational quality is a key consideration, a view echoed by the Incheon Declaration:

> An integral part of the right to education is ensuring that education is of sufficient quality to lead to relevant, equitable and effective learning outcomes at all levels and in all settings. Quality education necessitates, at a minimum, that learners develop foundational literacy and numeracy skills as building blocks for further learning, as well as higher-order skills.[54]

In this regard, ICTs play a vital role, and "must be harnessed to strengthen education systems, knowledge dissemination, information access, quality and effective learning, and more effective service provision".[55] In order to harness ICTs effectively, the Framework for Action argues for "[p]rovid[ing] teachers with adequate technological skills to manage ICT and social networks, as well as with media literacy and source criticism skills".[56] This is a facet we will explore in the section relating to ICTs for African education, and one of the examples used in the section relating to the future advancement of SDG 4 via space will relate specifically to the link between ICTs and the Framework for Action's call to "[r]eview, analyse and improve the quality of teacher training (pre-service and in-service) and

---

[51]World Education Forum, *Incheon Declaration and Framework for Action for the implementation of Sustainable Development Goal 4: Ensure inclusive and equitable quality education and promote lifelong learning opportunities for all*, 2015, 26, http://unesdoc.unesco.org/images/0024/002456/245656e.pdf (accessed August 16, 2018).
[52]Ibid.
[53]Ibid., 41.
[54]Ibid., 30.
[55]Ibid., 8.
[56]Ibid., 55.

[to] provide all teachers with quality pre-service education and continuous professional development and support".[57] First however, a broader consideration of quality education is needed.

2.2 Quality Education

At this point, two questions need to be asked in the context of SDG 4, Agenda 2063, and the Incheon Declaration. The first of these concerns what 'quality' education means, as called for in SDG 4 target 3, Agenda 2063 aspiration 1 target 7, and the Framework for Action, and the second concerns what SD means in the context of education since there are two aspects to consider here, namely SD *in* education (as alluded to in SDG 4.7) and SD *for* education. This is not a trivial distinction since education is called on to promote a culture of SD, but in order to successfully do so, education must *itself* be placed on a SD trajectory. We will explore what this entails in following sections, but it relies heavily on the use of the enabling technologies mentioned in SDG 17's target 8, and the ICTs mentioned in the Incheon Declaration.

In relation to quality education, it is important to remember that the "notion of quality is not a simple one; rather it is problematic, contested and multidimensional".[58] Critically, Southgate, Grimes, and Cox argue that:

> There is nothing natural about the concept of 'quality'. In higher education, 'quality; is a social construct that shifts across time and place and according to particular structural and historical forces'.[59]

A good starting point is thus to examine what students themselves regard as quality. Four themes were identified by Hill, Lomas, and MacGregor that point us to a constructive engagement with the notion of quality. First, the quality of the teacher or lecturer is vital, in the sense that they are knowledgeable, organised, and interested in students' input.[60] Related to this are the following: the flexibility of the lecturer in their delivery of the course, and their sympathy for the individual needs of the students; their relationship with the students and whether they are easy-going and helpful; and their responsiveness and trustworthiness.[61] Lecturer quality thus goes beyond being knowledgeable in a given subject and imparting that knowledge,

---

[57]Ibid.

[58]Yvonne Hill, Laurie Lomas, and Janet MacGregor, "Students' Perceptions of Quality in Higher Education," *Quality Assurance in Education* 11, no. 1 (2003): 17.

[59]Erica Southgate, Susan Grimes, and Jarrad Cox, "High status professions, their related degrees and the social construction of 'quality'," in *Achieving equity and quality in higher education: Global perspectives in an era of widening participation*, ed. Mahsood Shah and Jade McKay. (Cham, Switzerland: Palgrave MacMillan, 2018), 287–306.

[60]Hill, Lomas, and MacGregor, "Students' Perceptions of Quality in Higher Education," 16.

[61]Ibid., 16–18.

to "adequate and appropriate proficiency in organisation, presentation, assessment and evaluation".[62] It also relates to appropriate use of pedagogy to support student learning, as we will discuss later.

The second theme identified in relation to quality is student engagement with their learning, and whether they can relate to the curriculum and find it relevant for their experiences, as it expands their knowledge.[63] The third theme entails social and emotional support systems, whether the atmosphere is positive and values learning, and critically, whether small seminar groups and small group teaching is available.[64] In this last respect, it was noted that without access to such small group settings, vulnerable students who experience difficulties in adjusting to higher education could drop out, thus making it key to student survival. Indeed, in the context of distance education, it was found that "[i]ncreasing student-content interaction had the greatest effect [on student performance], followed by student-student interaction, with student-teacher interaction coming last".[65] Peer support is thus a vital facet of quality. The final theme is that of resources, such as libraries and Information Technology (IT).[66] It is this theme that SDGs 4A and 17.8 are addressing.

All four of these themes are touched upon in the Incheon Declaration, which notes that quality education requires

> relevant teaching and learning methods and content that meet the needs of all learners, taught by well-qualified, trained, adequately remunerated and motivated teachers, using appropriate pedagogical approaches and supported by appropriate information and communication technology (ICT), as well as the creation of safe, healthy, gender-responsive, inclusive and adequately resourced environments that facilitate learning.[67]

In relation to open, distance, and e-Learning (ODeL), to which we will turn later in the chapter, the quality of the final qualification must also be considered.[68] Here, in order to establish quality, ODeL must "provide a teaching and learning experience for students that is comparable with campus-based learning", with comparable performance in terms of outcomes, and a positive perception among "potential employers (or 'gatekeepers') about mainly online degrees in comparison with conventional degrees".[69] This is also reflected in Table 3, in the NDP's position that similar programmes in different institutions should provide a comparable education. These factors cannot be ignored in the educational quality equation, which summarises these facets in Fig. 1.

---

[62]Ibid., 18.
[63]Ibid., 17.
[64]Ibid., 18–19.
[65]Anne Gaskell and Roger Mills, "The quality and reputation of open, distance and e-learning: what are the challenges?," *Open Learning* 29, no. 3 (2014): 194.
[66]Hill, Lomas, and MacGregor, "Students' Perceptions of Quality in Higher Education," 17.
[67]World Education Forum, *Incheon Declaration and Framework for Action for the implementation of Sustainable Development Goal 4*, 30.
[68]Gaskell and Mills, "The quality and reputation of open, distance and e-learning," 193.
[69]Ibid., 198.

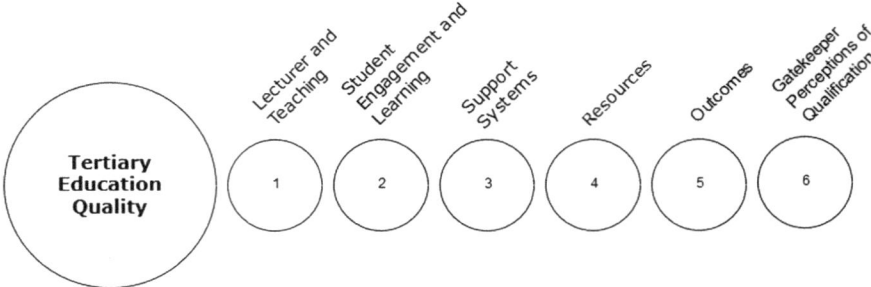

Fig. 1 Educational quality equation. (Based on the work of Yvonne Hill, Laurie Lomas, Janet MacGregor, and Anne Gaskell and Roger Mills.)

Ultimately, educational quality has a critical role to play in equity. While it is incumbent on higher education institutions to ensure equity in access, in relation to issues of gender and race (as in the case of South Africa for example), "it must be acknowledged that inequity in higher education has its roots in the lower levels of education", thus making it "essential to redress inequity at those levels and ensure quality education at both primary and secondary levels for equity measures at the level of higher education to be effective".[70] Issues of quality thus pertain to all levels of education, and without it equity is placed in jeopardy.

2.3   Sustainable Development and Education

When considering SD *in* education, SDG 4.7 points out that this entails knowledge and skills related to a culture of SD, which includes aspects such as global citizenship, peace, and diversity. However, a useful starting point for this discussion is the point raised by Kates, Parris, and Leiserowitz—in order to unpack SD itself, we must ask what is to be sustained, what is to be developed, and over what time period.[71] Since the SDGs relate to 2030, this is our guide in terms of time. However, while it is critical to develop skills and knowledge, Svanström, Lozano-García, and Rowe argue that since students will shape the future, it "implies rethinking and reorganizing higher education institutions (HEIs) to become effective change agents".[72] This makes it

> urgent to define new appropriate goals of higher education in terms of learning outcomes (LOs) for students, as well as generating the proper assessment criteria to check if the learning sought has occurred and create continuous improvement. And once we have done that, we must make sure that our organizational structures, program curricula, course

---

[70]Mohamedbhai, "Massification in Higher Education Institutions in Africa," 70.

[71]Kates, Parris, and Leiserowitz, "What Is Sustainable Development?," 3.

[72]Magdalena Svanström, Francisco J. Lozano-García, and Debra Rowe, "Learning outcomes for sustainable development in higher education," *International Journal of Sustainability in Higher Education* 9, no. 3 (2008): 340.

syllabi, and teaching and learning methods effectively address the LOs as well as other goals in HEIs.[73]

This is where the link can be made with the point that SD *for* education is an important consideration. It is not only the skills and knowledge of the educated that must be developed, but the learning environments and LOs themselves. This raises the question, what are appropriate LOs to promote SD?[74] Of course, since LOs encapsulate what it is that students should be able to do as a result of their learning activities, a "constructive alignment" is required so that "teaching and learning activities as well as student assessment are setup to make sure that and to test if LOs are achieved".[75]

Four facets in relation to LOs for SD are noted that move beyond development of skills and knowledge alone. The first is systemic or holistic thinking, which relates to the fact "that everything interacts with the things around it and that the world therefore consists of a complex web of relationships".[76] As such, we are reminded that *thinking itself* needs to be developed in line with SD, since sustainability renders it a necessity to "discern the patterns in a larger system and be able to understand cause-effect chains, understand conceptual models of a system, and create changes within and across systems".[77] The second aspect emphasises that education must promote the integration of different perspectives through inter- and multidisciplinarity because students must understand how their work will relate to both society and the environment on many levels including local and global, and in many contexts including political, social, and cultural.[78] Perhaps more than this, SD may ultimately require what Wallerstein terms *unidisciplinarity*, defined as the view "that in the social sciences at least, there exists today no sufficient intellectual reason to distinguish the separate disciplines at all, and that instead all work should be considered part of a single discipline".[79] While teaching still relies on disciplinary boundaries, perhaps such a grand unified perspective encapsulates best the need to look beyond these boundaries to see the world (including the social world) for the whole that it is. The third facet of LOs for SD is what SDG 4.7 referred to, namely skills and knowledge, such as "problem-solving, critical thinking, creative thinking, self-learning and skills related to communication, teamwork and becoming an effective change agent to shift policies, practices and societal norms".[80] The fourth facet entails the "awareness, attitudes and values" required to achieve SD.[81] Together, these encapsulate the term "transformative learning", understood as the

---

[73]Ibid.
[74]Ibid.
[75]Ibid.
[76]Ibid., 341.
[77]Ibid., 342.
[78]Ibid.
[79]Immanuel Wallerstein, *World-Systems Analysis: An Introduction* (Durham: Duke University Press, 2004), 99.
[80]Svanström, Lozano-García, and Rowe, "Learning outcomes for sustainable development in higher education," 342.
[81]Ibid.

competence to integrate, connect, confront, and reconcile multiple ways of looking at the world and the need for students to be able to cope with uncertainty, poorly defined situations, and conflicting or at least diverging norms, values, interests and reality constructions. ... the need to be able to change or shift perspectives related to cultures, disciplines, geographical conditions and time frames. ... the ability to go from local to global considerations, from short-term to long-term, and to realize that the world has been, is, and will be changing over time, which changes the conditions for people of different generations.[82]

Coupled with the indicators listed in Table 1, this summarises what we need to develop through the constructive alignment of education, LOs and SD. What then do we need to sustain? While many responses can be proffered for SD as a whole, ranging from resources, biodiversity, ecosystems, the environment, up to the Earth itself, for education the stakes cannot be higher. Carl Sagan pointed out that all aspects of modern life powerfully depend on science, and that science itself is "more than a body of knowledge; it is a way of thinking".[83] It is precisely this way of thinking, of critically engaging with the world that education needs to nurture and *develop*, so that we as responsible citizens can *sustain* the ability to meaningfully influence and steer national policies and "make intelligent decisions in our own lives".[84] This is by no means automatic or inevitable. Sagan warned here (applicable to the world), of his

foreboding of an America in my children's or grandchildren's time ... when awesome technological powers are in the hands of a very few, and no one representing the public interest can even grasp the issues; when people have lost the ability to set their own agendas or knowledgably question those in authority; when clutching our crystals and nervously consulting our horoscopes, our critical faculties in decline, unable to distinguish between what feels good and what's true, we slide, almost without noticing, back into superstition and darkness.[85]

The next section will explore how the implementation of SDG 4 (and the extended goals related to it) is faring in Africa and what Africa's own Agenda 2063 sets out in this regard. The section after this will then extend the discussion by considering how space is supporting this implementation, while linking up with the themes that were discussed above.

3   Implementation of Education and African SDG Progress

In ascertaining the extent to which the education-related SDGs are implemented, a useful first step is to review the SDG progress reports of the UN Secretary-General. In 2017, it was reported that inclusive quality education for all requires "increasing

---

[82]Ibid., 343.
[83]Carl Sagan, *The Demon-Haunted World: Science as a Candle in the Dark* (New York: Ballantine Books, 1996), 25.
[84]Ibid., 7.
[85]Ibid., 25.

efforts", especially in sub-Saharan Africa.[86] It was observed that "lack of trained teachers and the poor condition of schools in many parts of the world are jeopardizing prospects for quality education for all", with "less than 40%" of schools in half of sub-Saharan African countries having access to computers and the Internet for teaching purposes.[87] No mention was made of progress in relation to tertiary education, vocational training, or education for SD. It was however noted that "in 9 of 24 sub-Saharan African countries ... with data, fewer than half of the students at the end of primary education had attained minimum proficiency levels in mathematics".[88] This remains of great concern since tertiary education relies on the foundational skills of primary and secondary education. It is thus of grave concern that the World Economic Forum's Global Competitiveness Report 2017–2018 ranked the quality of South Africa's education system 118th out of 137, and the quality of its math and science education 128th out of 137.[89] Many other African countries fare similarly poorly. Accordingly, the 2018 SDG progress report of the Secretary-General pointed out that "[c]ritical efforts are needed to improve the quality of education ... and more investments in education infrastructure are required, particularly in the least developed countries".[90] Globally, it was noted that "617 million, or 58%, of children and adolescents of primary and lower secondary school age worldwide are not achieving minimum proficiency levels in reading and mathematics", while in "sub-Saharan Africa, only 37% of primary schools, 52% of lower secondary schools and 55% of upper secondary schools have access to electricity".[91] In the same region, only 61% of primary school teachers were considered trained. Again, tertiary education, vocational training, and SD-related education were not mentioned. With grave concerns remaining at the primary level, it is understandable that higher education and vocational training would not receive much attention, but this silence is concerning. Moreover, the 2018 Sustainable Development Goals Report provides no in-depth discussion for goals 1–5, noting that "[t]he indicators presented are those for which sufficient data are available to provide an overview at the regional and global levels".[92] Thus, very little data are available with which to assess progress of SDG 4, somewhat troubling after three

---

[86]United Nations Economic and Social Council, *Progress towards the Sustainable Development Goals: Report of the Secretary-General*, May 11, 2017, 7, http://www.un.org/ga/search/view_doc. asp?symbol=E/2017/66&Lang=E (accessed August 16, 2018).
[87]Ibid., 7.
[88]Ibid.
[89]Klaus Schwab, ed., *The Global Competitiveness Report 2017–2018* (Geneva: World Economic Forum, 2017) 269. http://www3.weforum.org/docs/GCR2017-2018/05FullReport/TheGlobalCompetitiveness Report2017%E2%80%932018.pdf (accessed August 16, 2018).
[90]United Nations Economic and Social Council, *Progress towards the Sustainable Development Goals: Report of the Secretary-General*, May 10, 2018, 6–7, https://unstats.un.org/sdgs/files/ report/2018/secretary-general-sdg-report-2018–EN.pdf (accessed August 16, 2018).
[91]Ibid., 7.
[92]United Nations, *The Sustainable Development Goals Report 2018* (New York: United Nations Department of Economic and Social Affairs, 2018), 18. https://unstats.un.org/sdgs/files/report/ 2018/TheSustainableDevelopmentGoalsReport2018-EN.pdf (accessed 17 August, 2018).

years. For the purposes of our discussion, we will focus on Internet connectivity, tertiary education enrolment, and continental initiatives as presented in Agenda 2063.

3.1 African Connectivity and Tertiary Enrolment

Internet connectivity in Africa remains a major concern because progress in many spheres, including education, heavily depends on the Internet, which "has reached the point where it, too, has become a necessity of modern life".[93] It is thus not an understatement to argue that "[p]roviding high-quality access to the Internet, with all of its capabilities for information search, education, healthcare, and e-commerce, promises to kickstart development in many remote regions".[94] As such, it is concerning to note the 2018 SGD progress report of the Secretary-General observed, under SDG 17, that

> Despite the worldwide increase in fixed broadband subscriptions, access to high-speed connections remains largely unavailable in the developing countries. In 2016, high-speed fixed broadband penetration reached 6 per cent of the population in developing countries, compared with 24 per cent in developed countries. Limitations in the capacity and speed of fixed broadband connections will affect the quality and functionality of this development tool and widen the already existing inequalities.[95]

In order to illustrate the challenge facing Africa in this regard, data compiled by the World Bank relating to Internet penetration (individuals using the Internet as a percentage of population) on the continent are presented in Fig. 2. These figures are all for 2016, and are more conservative than those presented in other sources (for example the Miniwatts Marketing Group[96]). Despite relatively high Internet penetration in Northern and Southern Africa (as well as the Indian Ocean island states), much of Africa still has very little and intermittent access to the Internet, to say nothing of high-speed connections. The worst performers on Earth are Eritrea (1%), Somalia (2%), Guinea-Bissau (4%), Central African Republic (4%), and Niger (4%). Thirteen countries have an Internet penetration rate of less than 10%, and they are all African.

One of the major challenges faced by the continent is that 16 of its 54 states are landlocked, and a major "cost and availability gap" has been observed between landlocked and coastal countries.[97] It is especially problematic for landlocked countries to access offshore fibre optic cables, and "[o]n average, the prices charged for Internet data in landlocked Africa are more than twice the prices for coastal countries".[98] This is one of the potential areas for space to facilitate educational programme delivery by extending Internet access to the interior of Africa, as will be

---

[93]Ward A. Hanson, "Satellite Internet in the Mobile Age," *New Space* 4, no. 3 (2016): 139.
[94]Ibid.
[95]United Nations, *The Sustainable Development Goals Report 2018*, 18.
[96]See https://www.internetworldstats.com/stats1.htm
[97]Hanson, "Satellite Internet in the Mobile Age," 144.
[98]Ibid.

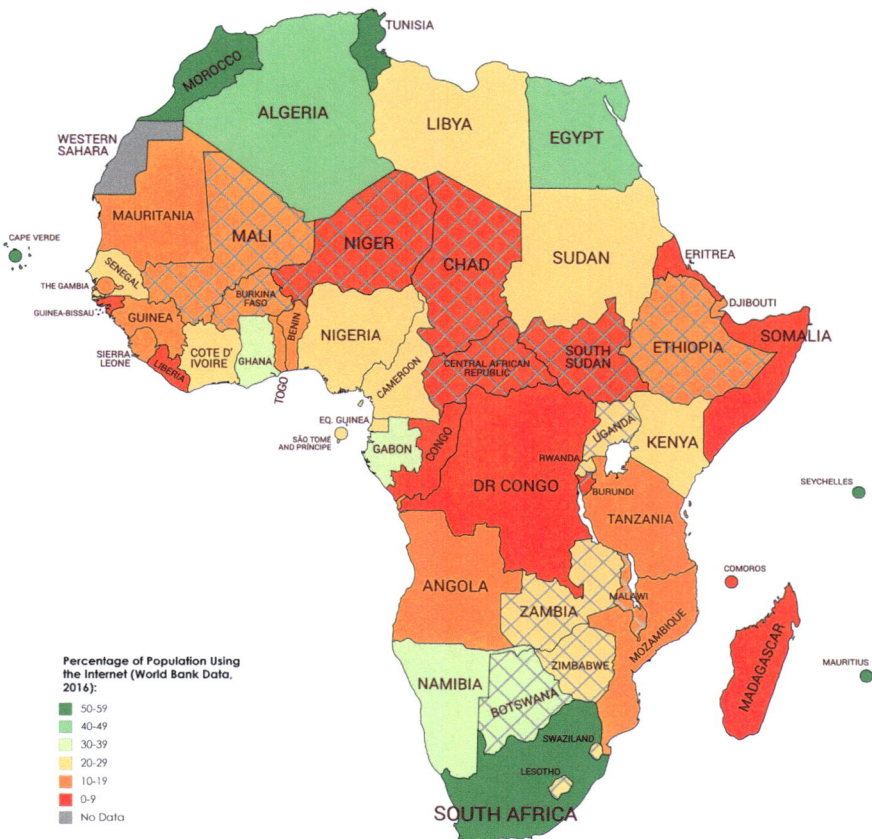

Fig. 2 African Internet penetration rate (and landlocked states). [The World Bank Group, *Individuals using the Internet* (% *of population*), 2018, https://data.worldbank.org/indicator/IT. NET.USER.ZS (accessed August 10, 2018). "Internet users are individuals who have used the Internet (from any location) in the last 3 months. The Internet can be used via a computer, mobile phone, personal digital assistant, games machine, digital TV etc.". Map created with mapchart.net (https://mapchart.net/). Landlocked countries indicated by crosshatch.]

explored later in this chapter. Figure 2 also indicates all landlocked countries in Africa, five of which have an Internet penetration rate of less than 10%.

While the overall rate of Internet penetration remains quite low, one of the best ways to address this challenge is through mobile phones. In contrast to Internet penetration, the overall rate of mobile phone penetration in Africa is relatively high, reaching a mobile cellular telephone subscription rate (both for prepaid and post-paid) of over 100 per 100 people in 18 countries. In this regard, the worst per-formers are Eritrea (10.21 mobile cellular subscriptions per 100 people), South Sudan (22 per 100 people), and Central African Republic (27.16 per 100 people). These data were also compiled by the World Bank, and are all for the same year as

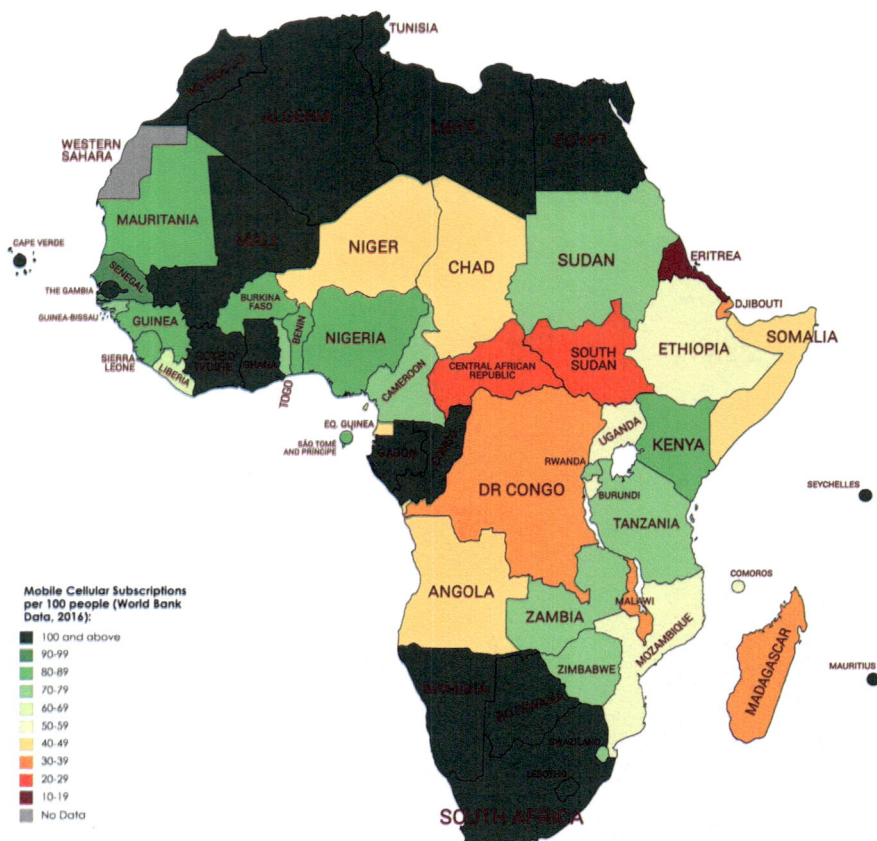

Fig. 3 Mobile cellular subscriptions per 100 people. [The World Bank Group, *Mobile cellular subscriptions (per 100 people)*, 2018, https://data.worldbank.org/indicator/IT.CEL.SETS.P2 (accessed August 13, 2018). "Mobile cellular telephone subscriptions are subscriptions to a public mobile telephone service that provide access to the PSTN using cellular technology. The indicator includes (and is split into) the number of postpaid subscriptions, and the number of active prepaid accounts (i.e. that have been used during the last three months). The indicator applies to all mobile cellular subscriptions that offer voice communications. It excludes subscriptions via data cards or USB modems, subscriptions to public mobile data services, private trunked mobile radio, telepoint, radio paging and telemetry services". Map created with mapchart.net (https://mapchart.net/).]

for the Internet penetration data above (2016), and are presented in Fig. 3. This provides a powerful comparative insight into how mobile phone use has proliferated in Africa and how it can continue to open the way for Internet access.

In total, there were about 557 million mobile users in Africa by 2017, of which feature phones made up 56% of the market share, and which "continue to outsell smartphones".[99] Feature phones can be defined as "a type of mobile phone that has more features than a standard cellphone but is not equivalent to a smartphone", being able to "make and receive calls, send text messages and provide some of the advanced features found on a smartphone".[100] Due to their cheaper costs, feature phones have been a driver of African mobile growth. While smartphone ownership data for African countries are scarcer than for mobile phones in general, the Pew Research Center has data for nine countries on the continent. Smartphone ownership rates for these, as a percentage of adults reporting ownership in 2015, are: South Africa (37%), Nigeria (28%), Kenya (26%), Ghana (21%), Senegal (19%), Burkina Faso (14%), Tanzania (11%), Uganda (4%), and Ethiopia (4%).[101] Undoubtedly, these rates have continued to increase since 2015. The use of mobile devices in general (feature phones, smartphones, and tablets) is a key component of educational programme delivery, and we will discuss this in depth later in this chapter, but it is worth pointing out that both the South African NDP and the Incheon Declaration acknowledged the importance of mobile devices for education, with the former noting that the "use of mobile devices such as phones and tablets in distributing learning content"[102] should continue to be explored, while the latter argued that "ICT, particularly mobile technology, holds great promise for accelerating progress towards this target [promoting numeracy and literacy]"[103] and that "the use of ICT, particularly mobile technology, for literacy and numeracy programmes" should be promoted.

At this point, it is useful to explore the extent of tertiary education enrolment in Africa, since this provides a clear understanding of the need for our focus on this sector. According to data again compiled by the World Bank, most African states have a gross tertiary enrolment percentage of less than 10% (see Fig. 4). While not all data depict the same year (for instance, the most recent data for Djibouti is from 2011, while the most recent data for Botswana is from 2017), it paints a very clear image of 'elite' higher education. Mohamedbhai summarises the work of Trow, who argued that in countries with national tertiary enrolment rates of up to 15% of the relevant age group, tertiary education constitutes "an elite system [that] caters to a privileged or talented group".[104] This is clearly the case for most of Africa, and even South Africa is close to falling in this category. The next category of countries, with tertiary enrolment rates of 16–50%, are argued to have 'mass' higher education

---

[99]Toby Shapshak, "Feature Phones Still Rule in Africa, As Smartphone Sales Slow," *Forbes*, March 28, 2017, https://www.forbes.com/sites/tobyshapshak/2017/03/28/feature-phones-still-rule-in-africa-as-smartphone-sales-slow/#4f9a740960e5 (accessed August 13, 2018).

[100]Techopedia Inc., *Feature Phone*, 2018, https://www.techopedia.com/definition/26221/feature-phone (accessed August 13, 2018).

[101]Jacob Poushter, "Smartphone Ownership and Internet Usage Continues to Climb in Emerging Economies," *Pew Research Center*, February 22, 2016, http://www.pewglobal.org/2016/02/22/smartphone-ownership-and-internet-usage-continues-to-climb-in-emerging-economies/ (accessed August 13, 2018).

[102]National Planning Commission, *The National Development Plan*, 303–304.

[103]World Education Forum, *Incheon Declaration and Framework for Action for the implementation of Sustainable Development Goal 4*, 47–48.

[104]Mohamedbhai, "Massification in Higher Education Institutions in Africa," p. 62.

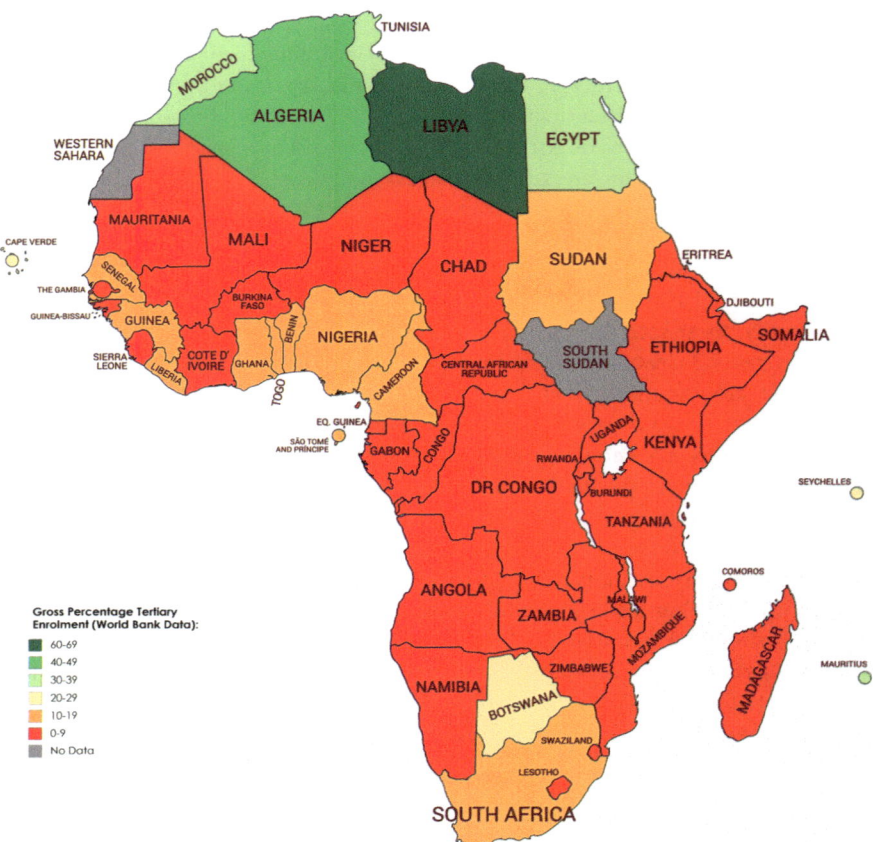

Fig. 4 African gross percentage tertiary enrolment. [The World Bank Group, *School enrollment, tertiary (% gross)*, 2018, https://data.worldbank.org/indicator/SE.TER.ENRR, (accessed August 10, 2018). "Gross enrollment ratio is the ratio of total enrollment, regardless of age, to the population of the age group that officially corresponds to the level of education shown. Tertiary education, whether or not to an advanced research qualification, normally requires, as a minimum condition of admission, the successful completion of education at the secondary level". Map created with mapchart.net (https://mapchart.net/).]

systems, where "higher education [is commonly perceived] as a right for those who are qualified".[105] This thus describes all other African countries except Libya, which had an enrolment rate of 60.5% in 2003 (the latest year for which World Bank data are available). While this situation has undoubtedly deteriorated given the conflict there in recent years, it came close to having a 'universal' higher

---

[105]Ibid.

education system, where higher education is regarded as "society's obligation to the population".[106] Mohamedbhai augments this scheme by arguing that only once 60% enrolment has been achieved can massification be said to have been attained.[107] However, we argue that this enrolment rate should be maintained throughout the period of study to ensure that graduation rates also reflect this massification.

This data thus reveal two facets of the post-secondary education landscape in Africa: first, it is mostly the preserve of the 'elites' who are fortunate enough to have access to it or the ability to study abroad, and second, the sector undoubtedly possesses immense growth potential, and it is inevitable that demand for higher education will continue to skyrocket, and without which SD and general economic growth will suffer, since "[t]here is a close relationship between a country's economic development and the proportion of its skilled workforce".[108] It is also likely that the kind of student protests for access to higher education that were seen in South Africa will become more commonplace across the continent if the demand is not met. Solutions are thus desperately needed that will allow for massification without undermining quality, while supporting improvements in graduation rates, and avoiding the pitfalls of massification experienced thus far: "[m]assification has had negative consequences on almost all of the public higher education institutions in Africa, including physical infrastructure, staffing, educational quality, graduate employment, and student mobility".[109] A related challenge is also the shortage of teachers, especially qualified ones.[110]

While the SDG progress reports are scant on details regarding tertiary education, Agenda 2063 sets out clear continental goals in this regard, which include the African E-University and the Pan African University (PAU) to help cater to growing educational demands. It is in this context that we will discuss ICTs for African education next.

3.2 ICTs for African Education

The infusion of ICTs into education is not a new phenomenon in African education, but there are many ways in which this can be enhanced and further promoted. Recent calls, such as that contained within the Incheon Declaration, to promote the use of ICTs, must take cognizance of both the history of past efforts and the remaining challenges and barriers that can hamper the use of ICTs for education in Africa. Because we will be focusing specifically on e-learning here, it is first necessary to define this concept. In our view, e-learning can be defined here as the

---

[106]Ibid.

[107]Ibid., 64.

[108]Ibid., 65.

[109]Ibid., 72.

[110]Masibulele Lunika, "Africa Must Invest in Remote Digital Education," *ITWebAfrica*, September 5, 2017, http://www.itwebafrica.com/ict-and-governance/523-africa/239979-technology-can-solve-africas-education-crises (accessed August 6, 2018).

use of educational technologies which are time and space independent, including applications used to design, deliver, implement, and evaluate multimedia content, such as assessments, discussions, and communications to cater for learning and teaching to both students and academics.

In 2006, it was noted that the African educational technology environment faced challenges related to lack of "robust telecommunications infrastructure", weak regulatory frameworks, low Internet penetration, lack of skilled ICT technicians, and poor maintenance.[111] More than a decade later, these same challenges—and others related to, for example, conflicts and lack of capital and investment—still confront the continent. It was also observed in 2006 that Africa's isolated e-learning initiatives had not yet coalesced into a broader network of "sustainable services" across the continent. This is also still true today. Some of the collaborative e-learning efforts that were observed during that time included the following, supplemented by more recent endeavours.

The first was the African Virtual University (AVU), based in Kenya, which was launched as a World Bank project in 1997 and after its transfer to Kenya was established as an intergovernmental organisation (IGO) in 2003 with 19 countries having signed the charter establishing the AVU as an IGO.[112] Having trained 74,073 students since 1997, and having over 53 partner institutions in 27 countries, the AVU is the leading pan-African e-learning network, covering Anglophone, Francophone, and Lusophone countries.[113]

As was noted above with the Agenda 2063 goals, it is aimed to create an E-University by 2023, and it was reported in 2017 that the African Union "will convert the existing African Virtual University … into an arm of the PAU, making it an Africa-wide university accessible to interested learners from across the continent".[114] The Pan African University was officially launched in 2011,[115] and operates under the vision to "develop institutions of excellence in science, technology, innovation, social sciences and governance, which would constitute the bedrock for an African pool of higher education and research".[116] Its particular focus is on the development of post-graduates, Ph.D.-candidates, and applied

---

[111]Daniel Gelaw Alemneh and Samantha Kelly Hastings, "Developing the ICT Infrastructure for Africa: Overview of Barriers to Harnessing the Full Power of the Internet," *Journal of Education for Library and Information Science* 47, no. 1 (2006): 12.

[112]Benin, Burkina Faso, Cape Verde Cote d'Ivoire, Democratic Republic of Congo, Ghana, Guinea, Guinea Bissau, Kenya, Mali, Mauritania, Mozambique, Niger, Nigeria, Senegal, South Sudan, Sudan, Tanzania, and The Gambia. African Virtual University, *AVU at a Glance*, 2018, http://www.avu.org/avuweb/en/avu-at-a-glance/ (accessed August 16, 2018).

[113]Ibid.

[114]Maina Waruru, "Pan African University to offer virtual education," *The Pie News*, November 16, 2017, https://thepienews.com/news/pan-african-university-to-have-an-odel-wing/ (accessed August 16, 2018).

[115]African Union, *Media advisory: Official launching of the Pan African University*, 2011, https://au.int/sites/default/files/newsevents/mediaadvisories/27573-ma-media_advisory_for_pau_0.pdf (accessed August 13, 2018).

[116]Pan African University, *Our Mission*, 2018, https://pau-au.net/about-us/our-mission/ (accessed August 13, 2018).

research.[117] Its de-centralised structure caters to all five sub-regions of Africa, with the sixth institute of PAU being the Pan-African E-University.[118] The various institutes of the PAU are: PAU Institute for Water and Energy Sciences (including Climate Change) (PAUWES) at the Abou Bekr Belkaid University of Tlemcen, Algeria (Northern Africa); PAU Institute for Life and Earth Sciences (including Health and Agriculture) (PAULESI) at the University of Ibadan, Nigeria (Western Africa); PAU Institute for Governance, Humanities and Social Sciences (PAUGHSS) at the University of Yaoundé, Cameroon (Central Africa); and the PAU Institute for Basic Sciences, Technology and Innovation (PAUSTI) at the Jomo Kenyatta University of Agriculture and Technology, Nairobi, Kenya (Eastern Africa).[119] The fifth institute is in the process of being established in South Africa, and will cater to Space Sciences. This critical development will be discussed shortly. The sixth institute, the Pan-African E-University is housed in Cameroon at the PAU Rectorate with the plan to "provide e-learning programmes to a wider audience of students and researchers".[120] How the existing AVU will relate to this is not yet clear, but building on the success of the AVU can raise the profile of ODeL as a viable study option for Africa in the future. The PAU reports that it has enrolled 1015 Master students and 308 Ph.D. students from all over Africa, with 364 graduates from 33 African countries.[121]

Another e-learning initiative with a relatively long history is eLearning Africa, an annual conference which draws "high-level policy and decision makers and practitioners from education, business and government",[122] and since the first conference in 2006, it has hosted 16,228 participants with over 85% coming from Africa, and over 3300 speakers focusing on "every aspect of technology supported learning and skills development".[123] It is also the venue for the Ministerial Round Table, where African ICT and Education ministers discuss issues related to education and technology.

A more recent African e-learning development was the establishment of e/merge Africa in 2014. It constitutes "a new educational technology network which is mostly for educational technology researchers and practitioners in African higher education" and offers online seminars, workshops, and short courses.[124]

---

[117]Deutscher Akademischer Austauschdiest, *Pan African University* (*PAU*), https://www.daad.de/miniwebs/ictunis/fr/29464/index.html (accessed August 16, 2018).

[118]Pan African University, *About Us*, 2018, https://pau-au.net/about-us/ (accessed August 13, 2018).

[119]Ibid.

[120]Ibid.

[121]Ibid.

[122]eLearning Africa, *Who attends?*, https://www.elearning-africa.com/conference_who.php (accessed August 16, 2018).

[123]eLearning Africa, *About*, https://www.elearning-africa.com/conference.php (accessed August 16, 2018).

[124]e/merge Africa, About e/merge Africa, https://emergeafrica.net/about-emerge-africa/ (accessed August 18, 2018).

It is also noteworthy that two of the 12 Agenda 2063 flagship projects directly relate to e-learning. These are the abovementioned AVU and E-University, as well as the Pan-African E-Network:

> An African Virtual and E-University. Increasing access to tertiary and continuing education in Africa by reaching large numbers of students and professionals in multiple sites simultaneously and developing relevant and high quality Open, Distance and eLearning (ODeL) resources to offer the prospective student a guaranteed access to the University from anywhere in the world and anytime (24 hours a day, 7 days a week).[125]

> The Pan-African E-Network. This involves a wide range of stakeholders and envisages putting in place policies and strategies that will lead to transformative e-applications and services in Africa; especially the intra-African broad band terrestrial infrastructure; and cyber security, making the information revolution the basis for service delivery in the bio and nanotechnology industries and ultimately transform Africa into an e-Society.[126]

Based on all these initiatives, we argue that ODeL is a very promising avenue for addressing the educational demands and challenges on the continent, albeit not without challenges of its own, such as poor Internet connectivity and the digital divide, which in the context of Africa "is so severe that the digital divide has been called the digital apartheid in reference to Africa, due to this continent's systematic exclusion from technological progress and its accompanying benefits" and on a domestic level "with African cities having higher levels of ICT development and access than African rural areas".[127] It is here we envision that space can have one of the most powerful roles supporting the educational SDGs, Agenda 2063 targets, and national educational goals, by facilitating ODeL programme delivery to areas lacking both telecommunications and educational infrastructure, in line with the observation that "Africa has been a fertile ground for the growth of open universities, independent distance education institutions, and a variety of technologically based programs and projects within universities", since the "issue then is not so much whether, but why it is necessary or a priority to invest in using ICTs in African higher education to meet the educational needs of its population and help solve educational challenges".[128] Because of this, we echo the argument that ICTs should continue to be infused into African higher education without delay, to increase productivity and efficiency despite (and because) of challenges related to educational resources on the continent. Some of the latest efforts to do just that across Africa are worth examining here in this regard.

In relation to Somali higher education, a study was conducted that focused on the University of Hargeisa, where the focus was on the utilisation of e-learning and

---

[125]African Union Commission, *Agenda 2063: The Future We Want—First Ten-Year Implementation Plan 2014–2023,* 17.

[126]Ibid., 17–18.

[127]Kwame Rivers, Patrick A. Rivers, and Vanessa Hazell, "Africa and Technology in Higher Education: Trends, Challenges, and Promise," *International Journal for Innovation Education and Research* 3, no. 5 (2015): 15.

[128]Ibid., 17.

to pinpoint factors relating to students' adoption of e-learning.[129] It was determined that one of the key priorities for Somalia to move forward in terms of its political and socio-economic transformation is to have a "sustainable education sector".[130] In line with our argument, Somalia has attempted to grapple with massification of higher education by "integration of technology enhanced education delivery".[131] This took place within the context of a society grappling with a digital divide since most ICT infrastructure is concentrated in urban areas, and even there most educational institutions use ICT for administrative purposes rather than for educational purposes. Even at tertiary level where ICTs are more commonly used or educational purposes, there is insufficient integration into wider educational system.[132] While some major university campuses in Somalia are connected to the Internet via local network providers, others have turned to space to seek solutions. The importance of e-learning for education, and for SD, is encapsulated in this Somali study by the view

> that e-learning in combination with easy to use ICT tools, access to OER [open education resources], and robust ICT support enhance the students['] educational experience and facilitate communication with the teachers, other students, and with the global community. Moving from classroom learning to e-learning has urged the students to shift in learning strategies which promoted their information literacy, computer skills, and active learning processes. In the light of future SDG, it is urgent to scale up the use of e-learning in post conflict setting such as Somaliland.[133]

In Kenya, a study investigating the educational barriers related to e-learning implementation for learning and teaching at three public universities found that the most serious barriers to scaling up the use of e-learning were (i) inadequate ICT and e-learning infrastructure, (ii) financial constraints, (iii) lack of operational e-learning policies, and (iv) lack of affordable and adequate Internet bandwidth.[134] It was recommended to address these challenges in order to successfully implement e-learning.

A Ghanaian study made the point that "despite these challenges some universities in Ghana are progressing beyond simple technology applications to leveraging and embedding various sources of online pedagogy, social, and collaborative

---

[129]Mohammed Omer, Tina Klomsri, Matti Tedre, Iskra Popova, Marie Klingberg-Allvin, and Fatumo Osman, "E-learning Opens Door to the Global Community: Novice Users' Experiences of E-learning in a Somali University," *MERLOT Journal of Online Learning and Teaching* 11, no. 2 (2015): 267–279.

[130]Ibid., 269.

[131]Ibid.

[132]Ibid.

[133]Ibid., 277.

[134]John K. Tarus, David Gichoya, and Alex Muumbo, "Challenges of Implementing E-Learning in Kenya: A Case of Kenyan Public Universities," *International Review of Research in Open and Distributed Learning* 16, no. 1 (2015): 129

Internet learning solutions for learning and educational purposes".[135] However, challenges again encountered include the following: not complying with national ICT policies, not expanding beyond the pilot phase of projects, lack of skills and training, inadequate ICT infrastructure, poor Internet, intermittent electricity, lack of expertise in e-pedagogy and instructional design, misalignment of technology with existing curricula, and inability to meet the pedagogical expectations of students and teachers.[136] It was found that the integration of ICTs into the curriculum is still the "formative stage" in Ghana (and we can add in most of Africa), thus it was determined that despite "modest overall performance", it cannot be said that e-learning implementation in Africa has 'arrived'.[137]

Most recently, a positive development was observed in relation to the "thriving ICTs" in Malawi, and at Mzuzu University specifically.[138] This university has invested heavily in ICTs in recent years, as lecturers have added mobile technologies to online distance learning and campus-based programmes in line with the national uptake of these technologies in Malawi. This development was enabled by the widespread availability and ownership of Internet-enabled devices including smartphones, as was observed in Fig. 3. This can serve as an example to other African countries with even higher mobile penetration. It was found that despite challenges, Malawi is ready to introduce a fully digital university that delivers courses "wholly via web-based technologies", and given the similar socio-economic conditions in many other African countries, that they too should be ready to embrace this concept.[139]

Another study, emanating from Tanzania, picking up this theme of the promise of mobile technologies, argued that while "access to computers and the Internet is still a challenge in many institutions in sub-Saharan Africa, the emergence of mobile devices brings a new hope".[140] It was accordingly argued that learning "[i]nstitutions should develop mobile interfaces that enable users to be able to access LMS [Learning Management System] via their mobile devices".[141] This is especially urgent since it was found when that study was done in 2015 that only 6% of African users browsing the Internet did so with desktop computers while 70% used mobile devices.[142]

---

[135]Josephine A. Larbi-Apau, Ingrid Guerra-Lopez, James L. Moseley, Timothy Spannaus, and Attila Yaprak, "Educational Technology-Related Performance of Teaching Faculty in Higher Education: Implications for eLearning Management," *Journal of Educational Technology Systems* 46, no. 1 (2017): 61–79.
[136]Ibid.
[137]Ibid.
[138]Paxton Zozie and Winner Dominic Chawinga, "Mapping an open digital university in Malawi: Implications for Africa," *Research in Comparative & International Education* 13, no. 1 (2018): 213.
[139]Ibid., 211–213.
[140]Joel S. Mtebe, "Learning Management System success: Increasing Learning Management System usage in higher education in sub-Saharan Africa," *International Journal of Education and Development using Information and Communication Technology (IJEDICT)* 11, no. 2 (2015): 57.
[141]Ibid., 58.
[142]Ibid.

Yet another study reflected on the increase of mobile learning research within Africa, climbing from three studies in 2010–11 to 13 studies 2014–16, revealing the growing interest in researching m-learning in African education.[143] This growing reliance on m-learning brings with it new needs related to provision of technical support to students related to the use of mobile technologies, the design of LMSs to be compatible with mobile devices (as argued by the previous study as well), the provision of training to course developers, and sufficient network access on campuses.[144]

This adaptation of LMSs should however not obscure the challenges facing the use of these systems in general in Africa. While many tertiary institutions have introduced LMSs, Mtebe argues that "the actual usage is reported low", that the "communication tools that are embedded in LMS such as discussion forums, chat, and e-mail are underutilized", that "many information systems implemented in developing countries tend to fail partially or totally", that in many sub-Saharan African institutions "users normally do not use the LMS after they have been trained", that many of the open source systems adopted in sub-Saharan Africa "suffer from usability problems", that it is "not uncommon to find many adopted LMS do not have enough quality learning materials uploaded in it", that sub-Saharan African users have insufficient exposure to information systems and that accordingly in these cases "confidence towards these systems is always low", that "many users cannot use LMS effectively due to lack of support services", and that "[m]any institutions in sub-Saharan Africa have either outdated [technology] policies or do not have such polices at all".[145] These are critical challenges to realising what we term sustainable e-learning (SeL), which is tied directly to SD.

3.3 Education for Sustainable Development and Sustainable e-Learning for Africa

To begin the discussion here, there is a need to reflect on how the UN itself views Education for Sustainable Development (ESD), and how our conceptualisation in this area moves beyond this. According to UNESCO, ESD is understood as "empower[ing] people to change the way they think and work towards a sustainable future".[146] Similarly, Kanbar defines ESD as a "vision of education that seeks to empower them [students] to assume responsibility for creating a sustainable

---

[143]Rogers Kaliisa and Michelle Picard, "A Systematic Review on Mobile Learning in Higher Education: The African Perspective," *TOJET: The Turkish Online Journal of Educational Technology* 16, no. 1 (2017): 11.

[144]Ibid.

[145]Mtebe, "Learning Management System success," 53–57.

[146]United Nations Educational, Scientific and Cultural Organisation (UNESCO), *Education for Sustainable Development*, https://en.unesco.org/themes/education-sustainable-development (accessed August 14, 2018).

future".[147] However, as we conceptualised and further expanded on this term, by distinguishing between SD *for* education and SD *in* education, ESD as defined here refers not to SD *for* education as argued by UNESCO, but to SD *in* education, since it refers to the content of the learning, and the LO's discussed earlier. Accordingly, we will use the term ESD+ to refer to the *method* of education, the form as it were (hence SD *for* education), in relation to effective learning environments to achieve the goals of SD *in* education (the original definition of ESD), thus placing education itself on a sustainable trajectory. In this regard, we introduce our concept of sustainable e-learning (SeL).

At present, the modern world is experiencing digital transformation as society makes way for the millennial generation. This dynamic process promises a fundamental change in all aspects of our lives, including knowledge dissemination, social interaction, business practices, political engagement, media, education, health, leisure, and entertainment.[148] Developing countries (especially in Africa) face numerous challenges in order to compete in the current global economy for SD. Even though the African continent is abundantly blessed with land for production of staple foods and mineral wealth, the development of these is commonly hampered by insufficient finances and, most importantly, human capital. Moreover, control over these resources is politically and financially contested by external actors. Thus, to increase the quality of human capital, nations are looking at both international and national investments (corporate), training, and education (in collaboration with HEIs), which we argue are fast becoming the norm to success for all parties involved. Practically, this is a time-consuming and at times overwhelming process, and especially financially constraining for developing countries, where educational systems are often feeble, underfunded, under-resourced, and are saddled with a lack of infrastructure, limited or no access to the Internet, formal education, and training, as was noted in the examples from literature used above.

Education is, in our view, the single most important factor and the central component of each country's approach to sustainable development. The role of education not only allows for knowledge creation and dissemination, but we firmly believe that it also allows for the creation of an 'inclusive society', allowing thus for equal opportunities for all within the educational biome. This reiterates the view that "good education should include socialisation where people can become part of existing socio-cultural, political and moral orders".[149] It thus means that "schools should engage socialisation in the form of citizenship education, character education and values education". Thus, we argue that the tertiary level presents a particularly

---

[147]Nancy Kanbar, "Can education for sustainable development address challenges in the Arab region? Examining business students' attitudes and competences on education for sustainable development: a case study from Lebanon," *Discourse and Communication for Sustainable Education* 3 (2012): 42.

[148]Marc Sehrt, "Digital divide into digital opportunities: e-learning in the Developing Countries," *UN Chronicle* 40, no. 4 (2003).

[149]Gert Biesta, "Good Education in an Age of Measurement: On the Need to Reconnect with the Question of Purpose in Education," *Educational Assessment, Evaluation and Accountability (formerly: Journal of Personnel Evaluation in Education)* 21, no 1. (2009): 33–46.

powerful opportunity to promote the concept of an inclusive society, since students have a sense of maturity that makes inducting them into communities of practice easier. An example to illustrate this is to have academic clubs at universities where the student members take the lead in terms of higher positions within organisations. Also, we argue for including and treating both undergraduates and postgraduates within departments as 'staff members' and on an equal basis (in the sense of involving them in research and imparting research skills, while also involving them in some institutional decisions). These initiatives at tertiary level lay the foundation for their future careers as they will know what is expected of them as graduates in society, and supports the NDP's goal of establishing "networks of excellence within and across institutions". As we have argued elsewhere, in this regard, Academic Developers can fulfil powerful roles within higher education by acting as 'hubs' "managing and fostering the information flow between lecturers, students, truly emerging eTools (such as Calculated Questions), and the broader ... learning community ecosystem".[150]

Over the past few decades across the globe the notion of e-learning in education has evolved alongside a new form of pedagogy to enhance traditional learning and teaching environments within HEIs, as demonstrated in the previous examples. In further deliberation of the definition of e-learning provided earlier, the concept of e-learning uses network technologies in order to design, develop, deliver, implement, and facilitate learning and teaching, in real and non-real time. Many advantages of e-learning exist, from providing cost-efficient learning experiences (students only requiring stable Internet connection and a mobile device), different training options and learning experiences, and participation within MOOCs (Massive Open Online Courses), thus giving students from developing countries a priceless experience of gaining educational experiences comparable to that of the developed world. However, e-learning is not only network dependent as there exist offline educational technologies commonly referred to as Personal Learning Environments (PLEs). PLEs have been recognised as "a key component in a distributed, connectivist, learning model".[151]

At this juncture, it is necessary to define and differentiate our concept of SeL from similar concepts used elsewhere. Existing definitions include that of Robertson, "e-learning that has become normative in meeting the needs of the present and future"[152]; the National Committee of Inquiry into Higher Education "the adoption of technology to maintain teaching quality at reduced unit costs"[153]; Arneberg et al. "programmes being offered on a continuous basis and not phased out after a defined

---

[150]Valentino Van de Heyde and André Siebrits, "Students' attitudes towards online pre-laboratory exercises for a physics extended curriculum programme," *Research in Science & Technological Education*, 2018: 22.

[151]Ibid., 7.

[152]Cited in Ahmed D. Alharthi and Maria Spichkova, "Individual and Social Requirement Aspects of Sustainable eLearning Systems," *International Conference on Engineering Education and Research*, Sydney, Australia (2016): 2.

[153]Cited by Littlejohn in Karen Stepanyan, Allison Littlejohn, and Anoush Margaryan, "Sustainable eLearning in a Changing Landscape: A Scoping Study (SeLScope)," Report Prepared by the *Higher Education Academy Supporting Sustainable eLearning Special Interest Group* (2010): 46.

project period or after specific subsidies are terminated"[154]; Meyer "policies and practices that improve the likelihood that an online educational program will be financially viable"[155]; Stepanjan, Littlejohn, and Margaryan "cannot be explored without consideration of the rapid and continual development of digital technologies. Technological affordances open up new, ubiquitous opportunities for people to learn in a number of ways using a variety of approaches"[156]; and Stepanjan, Littlejohn, and Margaryan "[s]ustainability is the property of e-learning practice that evidently addresses current educational needs and accommodates continuous adaptation to change, without outrunning its resource base or receding in effectiveness".[157] Another view on the achievement of sustainable e-learning, is that of Nichols,[158] who argues that "it is necessary to implement it strategically with clear and open communication channels, sufficient resources, targeted professional development, and a willingness to revise institutional systems so that e-learning 'fits' across the entire enterprise".

Our concept of SeL builds on, and goes beyond, these definitions, and instead combines aspects thereof with additional features. To make e-learning itself sustainable, and for it to support SD, SeL must: (i) be guided by an overarching institutional e-learning strategy, (ii) not be used in a silo in the sense of not being used in isolation, but utilised in the full capability, and integrated fully into academic discipline (in contrast to Mtebe's example of LMS usage discussed earlier), (iii) the setup of an e-learning environment should be constructed in a scaffolded manner (including signposts for users[159]), (iv) include up-to-date educational technologies for formative and summative assessment, (v) should promote and facilitate student engagement with the learning material and with each other, (vi) should be subject to continuous re-evaluation, including introduction of innovative e-pedagogy, and (vii) that Academic Developers should possess both subject knowledge and e-pedagogy skills. These components of SeL are depicted in Fig. 5. It is also necessary for African e-learning practitioners to be more open to continental collaboration, which could lead to international collaborations for the SD of the continent.

The need for SeL within higher education is encapsulated by the findings of Omer et al., in that

> institutions need to build capacity and structures to provide their students with technical and pedagogical support. The faculty staff needs to develop specific pedagogical skills adequate for using ICT based and self-directed learning that promote life-long learning. Community awareness of the use of ICT is needed to increase the social reinforcement of e-learning".[160]

---

[154]Ibid.

[155]Ibid.

[156]Ibid., 29.

[157]Ibid., 10.

[158]Mark Nichols, "Institutional Perspectives: The challenges of e-learning diffusion," *British Journal of Educational Technology* 39, no. 4 (2008): 607.

[159]Scaffolding is most prominently associated with the work of Gilly Salmon, see for example https://www.gillysalmon.com/five-stage-model.html

[160]Omer et al., "E-learning Opens Door to the Global Community," 277.

Fig. 5 Sustainable e-learning (SeL) model

In order to mitigate the challenges surrounding usage of an LMS (as summarised by Mtebe), academics in collaboration with academic developers should use the LMS effectively for SeL in terms of creating online learning content relevant to the topic, to move away from promoting Lower Order Thinking Skills (LOTS) to Higher Order Thinking Skills (HOTS)—in relation to Bloom's digital taxonomy[161]—and implement pre-requisites (for example pre-laboratory quizzes in order to gain entry into laboratory sessions) within the LMS in order for students to progress.

HEIs from developing countries should all consider evaluating the impact of their current learning and teaching practices across all disciplines, from the design and development to the final implementation. This would allow educational practices that would appeal to the most students from various educational backgrounds. However, in order for this to happen, numerous factors need to be taken into consideration, ranging from educational expertise, technological know-how, infrastructure, capital, etc. In addition, each department, faculty, and HEI within Africa should have their own e-learning strategy. This strategy should include an annual review to assess the educational technology implementation in all the courses, which will aid in the attitude to further expand (on the utilisation of these ICTs) on the reach of educational prospects in HEIs, thus supporting and contributing towards economic and social development, and embracing higher education and education in general in the development process. Therefore, to create an interconnected educational biome, allowing all inhabitants to participate in the decision making processes within this community, key and expert competences are needed. These educational competences will be discussed here.

In terms of sustainability, students must be taught and enabled to deal with interconnected, multifaceted difficulties throughout their university studies, but

---

[161]Lee Watanabe-Crockett, Bloom's Digital Taxonomy Verbs, *Global Digital Citizen Foundation*, June 11, 2015, https://globaldigitalcitizen.org/blooms-digital-taxonomy-verbs (accessed August 17, 2018).

most importantly during their working life. These are aligned with the LOs for SD discussed earlier. In order to move towards this realization, students must gain experience during their tertiary education. One example to illustrate this concept is to move science students away from the traditional based (cookbook method) laboratory sessions towards design based labs, which will contribute towards real-life situations. Some examples of this include searching for oil on university campuses as a practical exercise of applying the concept of gravity, determination of a toy car's spring constant utilising the concepts of Hooke's Law and Simple Harmonic Motion. This implies considering the educational biome as acting as one system, which in turn will act on another system, i.e. society, industry, etc. Overall, these students should acquire the sound knowledge and other related skills relevant to their own professional research field, which in turn allows for possible employment, and is therefore the prerequisite for involvement in society and industry. Students should also possess other skills, such as to be digitally competent in order to energetically and maturely shape present-day societies, thus moving beyond the professional skills needed for their careers, allowing therefore for wider visions for a sustainable society and the further development for education in SD.

Numerous studies have been done regarding what 'good education' is, but we argue that academics should use pedagogy that works for their educational environment. Developing countries cannot use pedagogies that have emerged from, and have a long history within, developed countries, as these are worlds apart in terms of infrastructure, social status, research, internet availability, digital devices, and other factors. Due to various pedagogies used, 'good education' should at least entail the use of educational learning technologies to enhance experiences for both academics and students, thus promoting or moving forward in terms of SeL within HEIs. Furthermore, this idea of 'good education' using educational learning technologies promotes and supports the current generation, i.e. millennial generation, that thrives on using the Internet and current electronic devices to communicate, view and share videos, and also to do their studies. This online environment (i.e. LMS) thus caters for the needs of the millennials and tries to warrant that learning using technology is inspired by the creativity and imagination of the students.

However, while SeL holds great promise for Africa's tertiary education sector, how these initiatives are evaluated over time, especially on the African continent, is critical to their effectiveness. The use of e-learning in Africa is dependent on successfully overcoming numerous barriers that hinder the use of technology within learning and teaching environments (such as discussed in previous examples). These factors range from (but are not limited to): disbelief of technology for educational benefit, academic comfort zones in the sense of using only traditional teaching methods, investment and support from government, stable internet, access to digital devices, educational and technical expertise, and lack of knowledge to design, develop, implement, and evaluate projects involving e-learning use. An added factor is the relatively high cost of Internet availability in developing countries, most often as a result of misguided telecommunication regulations and punitive taxation that discourage the development of internet access service and mobile networks through competition. Another challenge is that some academics are only using the most common functions which include dumping of course notes (without lack of appropriate signposting, for example file naming) and sending announcements, which may not be the most effective method of employing e-learning, as also argued by Mtebe.

The authors therefore argue that SeL is the approach that can possibly overcome many of the challenges faced by Africa and its underserved students. Thus, in an effort to unlock the full potential of SeL in relation to education, which in turn relates to SD, we suggest that systematic reviews and feasibility studies should be conducted in HEIs relating to implementation of e-learning. It has in the past been argued that higher education is often considered to be "an expensive and inefficient public service that largely benefited the wealthy and privileged", but now it is abundantly clear that it is vital for "successfully boosting a country's productivity, competitiveness, growth, innovation, and performance across key economic sectors".[162] We thus stand in solidarity with the views of Sehrt,[163] who argues that if education and capacity-building are critical steps for entering into the new global economy, the use of e-learning should be considered as a critical facet for basic SD, an alternative medium of capacity-building, and a means to empower people. It is also worth repeating Nhando's views here

> The goal of delivering a high quality education to every child in Africa remains unfulfilled, but technology presents an opportunity for this to be a reality. eLearning has overwhelming potential to improve education systems in African countries and if implemented well with strategies that focus on overcoming these key challenges, radical transformation of the education system is possible.[164]

Nhando also illustrates why our in-depth discussion of e-learning within the context of African higher education was necessary, and why efforts must continue to point out the ways in which technology can support learning

> Teachers on the continent have been brought up in education systems with limited technology and they find it difficult to utilize technology to engage and support learning. There is a great emphasis that needs to be made for teachers to understand that technology is not replacing them, but rather it is an enabler that will enhance their work.[165]

In the next two sections, we will explore how space has been supporting the implementation of SDG 4 in Africa, and how it can advance it further in future. One of the key factors in this discussion will be how space supports educational programme delivery (especially in relation to ODeL) via communication technologies.

---

[162]Diane E. Eynon, *Women, Economic Development, and Higher Education: Tools in the Reconstruction and Transformation of Post-Apartheid South Africa* (Cham, Switzerland: Palgrave Macmillan, 2017), 20.

[163]Sehrt, "Digital divide into digital opportunities."

[164]Danai Nhando, "3 Key Challenges of Implementing eLearning In Africa," *eLearning Industry*, October 30, 2015, https://elearningindustry.com/3-key-challenges-implementing-elearning-in-africa (accessed August 18, 2018).

[165]Ibid.

Expanding access to ODeL across Africa will only happen with the support of space-based technology, but this is only one factor in the complex symbiotic interrelationship between space and SDG 4.

4   The Role of Space in Implementing SDG 4 in Africa

The discussion here will be guided by our Symbiotic Education Enabler Model for Space (SEEMS) (Fig. 6). This model is based on three mutually reinforcing interrelationships between space and education—each forming in our view a 'pillar' of the African education-space ecosystem. The first of these is derived from UNOOSA's earlier assertion that space supports education in two ways—by facilitating educational programme delivery, and by motivating students to pursue studies in the natural sciences. This represents the most direct way in which space activities, applications, and data support education (including SDG 4). However, in addition to this it is necessary to recognise a specific subset of education, namely space-related education (astronomy, aerospace engineering, Earth and Planetary Sciences, space law, and others). This space education also supports the SDGs, including SDG 4, in a general way by promoting the acquisition of science, technology, engineering, and mathematics (STEM) skills, as well as technical and vocational skills. Not only are students thus motivated to study sciences, as UNOOSA argues, but many students are also motivated to study space studies because of the positive influence of space in supporting education. This thus forms

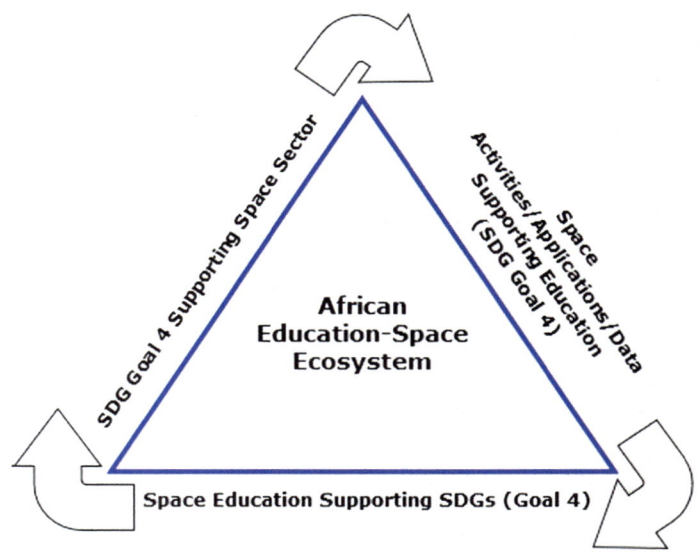

Fig. 6 The symbiotic education enabler model for space (SEEMS)

the second 'pillar' of the SEEMS model. Finally, these urgently needed skills then feed into the next generation of space professionals, thus highlighting that SDG 4 also supports space, just as space supports SDG 4. This space sector is precisely what produces the next cycle of support for education and motivation for studying space, thus producing an upward spiral. The SEEMS model will structure the discussion here, beginning with the first 'pillar'.

Before delving into the pillars however, it is instructive to identify to what extent the SDGs, Agenda 2063, and the NDP have identified and recognised the importance of space in supporting their goals. While space is absent from the SDGs themselves, there are some efforts to make linkages between space-related projects and the 2030 development agenda. One example is a UNOOSA and European GNSS Agency publication entitled *European Global Navigation Satellite System and Copernicus: Supporting the Sustainable Development Goals*.[166] That study "investigates how European Union space technologies support the fulfilment of the SDGs" and argues that "Earth observation (EO) and geolocation (provided by global navigation satellite systems (GNSS)) is recognized by the United Nations in supporting the achievement of the SDGs".[167] In relation to education, the publication indicates that

> The European GNSS Agency (GSA), signed an MoU with UNOOSA, to collaborate on educational events that could help build capacity in developing countries. In addition, other capacity-building initiatives are also in place or have been promoted in the past, including: the e-Knot project (to strengthen the interaction between education-research-industry in Europe), the BELS project (facilitating the breakthrough of EGNSS technology in South-East Asian industry), the SATSA project (offering training to the South African authorities on satellite navigation and augmentation systems) and the Space Expo exhibition (introducing space technology to the general public).[168]

A case study is identified where a "GNSS-based platform helps assess education facilities in Nigeria [and] [f]acility assessments or surveys can provide governments with valuable information to improve service delivery or to better plan future infrastructure investments", and it is identified that ICTs "offer unprecedented opportunities for using new types of complementary metrics and data".[169] There is thus in some instances an emerging awareness of how space can support the SDGs, but there is no integration of space into the SDGs themselves.

In relation to Agenda 2063, space has been integrated as a flagship programme.[170] In this regard, while space is not directly linked to the educational goals discussed earlier, a call is issued to "[d]evelop/implement frameworks for the

---

[166]United Nations Office for Outer Space Affairs (UNOOSA), *European Global Navigation Satellite System and Copernicus: Supporting the Sustainable Development Goals—Building Blocks towards the 2030 Agenda* (Vienna: United Nations, 2018). http://www.unoosa.org/res/oosadoc/data/documents/2018/stspace/stspace71_0_html/st_space_71E.pdf (accessed August 19, 2018).

[167]Ibid., 1.

[168]Ibid., 92.

[169]Ibid.

[170]African Union Commission, *Agenda 2063: The Future We Want—First Ten-Year Implementation Plan 2014–2023*, 18.

facilitation of adoption of curricula in space technology in member states universities/polytechniques" by 2023,[171] and Agenda 2063 incorporates the recognition that

> Outer space is of critical importance to the development of Africa in all fields: agriculture, disaster management, remote sensing, climate forecast, banking and finance, as well as defense and security. Africa's access to space technology products is no longer a matter of luxury and there is a need to speed up access to these technologies and products. New developments in satellite technologies make these very accessible to African countries.[172]

In the South African NDP, space is again not linked directly to education, but space science and technology is one of the programmes aimed at "stimulating research and innovation" under the auspices of the Department of Science and Technology.[173] Additionally, space and software engineering is listed as one of the areas in which South Africa has comparative and competitive advantages, and "[s] special consideration should also be given to dedicated programmes" in this area.[174] From these examples it thus becomes apparent that some appreciation exists on all three levels of the role of space in supporting developmental goals, but there is room for a much closer alignment between space and education. As such, the next section will look at the first 'pillar' of the SEEMS model, and explore some ways in which space directly supports education in Africa.

4.1 Space Supporting Education

Space technology, data, applications, and activities most directly impact education in the context of the previous discussion through expanding access to the Internet. For all its promise to expand access to education, e-learning requires access to the Internet. As we have discussed, African education systems will have to accommodate millions of new entrants by 2030, and in most cases existing institutions of higher education are struggling to accommodate current demands for access. As the NDP 2030 example showed, in South Africa there is a national imperative to dramatically increase university enrolments and graduation rates, and the reason why this example is so pertinent is because South African challenges relating to massification are not dissimilar from those encountered in other African countries. The example of one South African university, the University of the Western Cape, can serve to illustrate this. This university's Institutional Operating Plan 2016–2020 White Paper makes it clear that the institution has an "enrolment mandate from the Department of Higher Education and Training …require[ing] it to manage student enrolment and throughput in line with nationally agreed targets".[175] Related to this,

---

[171]Ibid., 92.
[172]Ibid., 18.
[173]National Planning Commission, *National Development Plan 2030*, 326.
[174]Ibid., 327.
[175]University of the Western Cape, *Institutional Operating Plan 2016–2020 White Paper*, 2016, 10, https://ikamva.uwc.ac.za/content/whitepaper.pdf (accessed August 18, 2018).

"the increase in residence places on campus has not kept pace with enrolment growth", such that in 2015 only 22% of students could find accommodation in a residence owned or leased by the institution.[176] Rapid expansions in quality ODeL options are thus required to expand access to education beyond what existing physical HEIs can accommodate. As Figs. 2 and 3 showed, mobile devices, specifically phones, have proliferated across the continent, but quality and affordable Internet access remains a major problem.

One way to connect existing university campuses with the Internet is via satellite. In the case of Somalia, one of the countries with the lowest Internet penetration in the world, it was observed that "[s]ome of the universities have VSAT [very small aperture terminal] dish installed on the campus, which is connected directly to international satellite-based ISP [Internet Service Providers]".[177] VSAT "offers internet connectivity and voice, video and IP solutions on a shared bi-directional satellite platform"[178] and provides "global connectivity coverage through a network of satellites".[179] VSAT is also used by industries that are operating beyond the easy reach of existing terrestrial copper or fibre networks, such as the mining, oil and gas, and maritime industries. On top of existing VSAT services, at least eight new satellite constellations are being planned in Low Earth Orbit (LEO) and Medium Earth Orbit (MEO) to expand internet access, beginning deployment in 2018.[180] These are OneWeb, Telesat LEO, SES O3B, Iridium Next, LeoSat, SpaceX, Samsung and Boeing. Other contenders include Kepler Communications, Telesat Canada, Theia Holdings A, Inc., Spire Global, ViaSat, Karousel LLC, Audacy Communications, and Space Norway AS.[181] Among only Samsung and Boeing there are plans to deploy over 7000 new satellites to provide Internet access (4600 for Samsung to LEO, and 2956 for Boeing).[182] In the case of Samsung, this will provide a 200 gigabyte per month service "for up to 5 billion users".[183] SpaceX has already launched two demo satellites in February of this year,[184] and their constellation (Starlink) is planned to consist of 4425 satellites[185] with an additional 7518 satellites in another SpaceX constellation.[186] Meanwhile,

---

[176]Ibid., 47.

[177]Omer et al., "E-learning Opens Door to the Global Community," 269.

[178]Sentech, *VSAT*, 2016, http://www.sentech.co.za/content/vsat (Accessed August 18, 2018).

[179]Internet Solutions, *VSAT Satellite connectivity to extend your business reach*, 2018, https://www.is.co.za/solution/vsat/ (accessed August 18, 2018).

[180]Matt Williams, "This is the Year Internet from Space Gets Really Serious," *Universe Today*, January 8, 2018, https://www.universetoday.com/138210/year-internet-space-gets-really-serious/ (accessed August 18, 2018).

[181]Douglas Messier, "SpaceX to Launch Global Satellite Broadband Test Spacecraft on Wednesday," *Parabolic Arc*, February 18, 2018, http://www.parabolicarc.com/tag/space-norway/ (accessed August 18, 2018).

[182]Ibid.

[183]Ibid.

[184]Mike Wall, "SpaceX's Prototype Internet Satellites Are Up and Running," *Space.com*, February 22, 2018, https://www.space.com/39785-spacex-internet-satellites-starlink-constellation.html (accessed August 18, 2018).

[185]Williams, "This is the Year Internet from Space Gets Really Serious."

[186]Messier, "SpaceX to Launch Global Satellite Broadband Test Spacecraft on Wednesday."

OneWeb has set the goal of achieving "Gigabit per second speeds, lower latencies, and affordable self-installed terminals" to "support both our 2022 goal of connecting every unconnected school and our 2027 goal of bridging the digital divide".[187] With all these ambitious endeavours taking place the current poor connectivity in Africa is set to change rapidly, as long as the services are affordable to poor communities. It also promises to make Internet access cheaper for the many landlocked countries in Africa and elsewhere.

As mentioned previously, the use and stability of the Internet remains crucial for ESD+ in Africa and the implementation of ICTs within learning and teaching environments, however there are alternatives in terms of offline educational ICTs (which do not need the Internet for functionality). On the other hand, the user will need an Internet connection to download these educational applications, but this will be the only time the Internet will be used. A reliable Internet connection within actual classrooms is an enjoyable contributing factor towards improving the learning and teaching for both students and academics, but offline tools can still provide the users (students and academics) with the ability to portray the role of 'a creator and receiver of resources'. Furthermore, the utilisation of offline tools, "aims to move learners from a traditional classroom environment to a group conversation based on learning" allowing the students with "mobile and personal tools that enable them to work offline".[188] In other words, these technologies can consequently enhance the educational experience and should be an add-on for both offline and face-to-face learning and teaching activities, thus providing and reaching deserving students who can also use educational technologies (albeit offline) to enhance their learning experience.

The Indian example of EDUSAT, the world's first education satellite which launched in 2004, designed to "use the virtual classroom concept to offer education to children in remote villages, quality higher education to students in areas without access to good technical institutes, adult literacy programmes and training modules for teachers" offers a pertinent case study for leveraging these satellite constellations for African education.[189] Based on the EDUSAT programme, India rolled out both Satellite Interactive Terminals and Receive Only Terminals to implement 83 networks connecting over 56,000 schools and colleges across the country by 2012.[190] This case represents an example of the potential of modern tele-education, incorporating online lectures and exams, continuous education programmes,

---

[187]Greg Wyler, "We All Need Access," OneWeb, 2018, http://www.oneweb.world/ (Accessed August 18, 2018).
[188]Mafawez T. Alharbi, Amelia Platt, Ali H. Al-Bayatti, "Personal Learning Environment," *International Journal for e-Learning Security (IJeLS)* 3, Issues 1 (2013): 281.
[189]Padma Tata, "India launches world's first education satellite," *New Scientist Ltd.*, September 20, 2004, https://www.newscientist.com/article/dn6423-india-launches-worlds-first-education-satellite/ (accessed August 19, 2018).
[190]Indian Space Research Organisation (ISRO), *Tele-Education*, 2017, https://www.isro.gov.in/applications/tele-education (accessed August 19, 2018).

libraries, databases and more. As we discussed with SeL, African educators will also have to be ready to meet the challenge of expanding e-learning across the continent.

Kasturirangan summarises the above neatly by pointing out that "[s]pace has enabled the information and communications technology (ICT) era and brought in the concept of the 'global village'; it is now becoming a tool for supporting education", with the focus on

> imparting quality education to students using satellites—thus bridging the gap of taking quality education opportunities to different parts of the country [in India]. The space system has the ability to provide instant and continuous connectivity between places where one end has the best of teachers and teaching materials and on the other side are students that can absorb these teaching systems spread across virtual class-rooms.[191]

Developing countries must thus take cognizance of the Indian example, which illustrates that

> space has proven capabilities of being able to bridge communities in the fields of quality education and health-care; it has great potential to address the looming problems of illiteracy, sub-optimal education, lack of health-facilities, access to quality health-care, etc.[192]

Of course, all of this depends, as Nhando points out, on the "the level of investment needed in developing local content that is aligned with national curriculums and that can be utilized for eLearning",[193] which is why we focused on the SeL concept.

While the above discussion of Internet provision relates to UNOOSA's example of supporting educational programme delivery, an example from Uganda clearly demonstrates the ability of space knowledge to act as a motivator to "help students to discover new interests".[194] We have already argued that tertiary education relies to a large extent on the foundational skills established at primary and secondary levels, making the Ugandan example a showcase for other African countries. There, the National Curriculum Development Centre (NCDC) reported that "it will integrate the study of earth and space science in the revised curriculum for secondary schools", thereby "incit[ing] the students to look for opportunities".[195] While it is unclear how far this effort has progressed, it is a clear step in the right direction in terms of introducing foundational space knowledge into secondary schools, thereby motivating students to pursue further studies in science, and possibly even space sciences. The formation of Astronomy clubs at university level, such as the

---

[191]Krishnaswamy Kasturirangan, "Space technology for humanity: A profile for the coming 50 years," *Space Policy* 23 (2007): 160–161.

[192]Ibid., 165.

[193]Nhando, "3 Key Challenges of Implementing eLearning In Africa."

[194]Patience Ahimbisibwe, "Secondary students to study space science," *Daily Monitor*, May 13, 2016, http://www.monitor.co.ug/News/National/Secondary-students-to-study-space-science-/688334-2714476-6fdvbg/index.html (accessed August 18, 2018).

[195]Ibid.

University of the Western Cape and Cape Peninsula University of Technology Student Space Association,[196] can further help to promote communities of practice and expand space knowledge to all interested individuals.

4.2 Space Education Supporting SDG 4

We have seen how the SDGs call for relevant technical and vocational skills, vocational training, and science and technological innovation. This section will consider how space education specifically helps to support these goals, in the second 'pillar' of the SEEMS model. Across Africa, countries are gearing up new space programmes and acquiring their first satellites. For these countries, there have been a few options in this regard. As Wood and Weigel argue, "[t]here are many ways a country can achieve this milestone [of their first satellite in LEO]", and options include "design and build the satellite locally", "produce the satellite in partnership with another country", and "buy the satellite from a foreign company".[197] These options, and others, are incorporated into Wood and Weigel's Space Technology Ladder, but the particular milestone of concern for the discussion here is procuring a LEO satellite with training services.[198] Through the Space Technology Ladder it becomes possible to track the milestones made by countries as they progress up to more complex space activities. Most African countries are still in the early stages of the ladder, but this aspect of purchasing a satellite with training services is a potent example of vocational training serving the SDGs. For the most part, when a country decides to procure its first satellite, skills to build and operate it locally are not available or sufficient, and in this case countries turn to training packages. Algeria serves as a good example here. In several different instances Algeria has procured or jointly built a satellite with foreign partners, but in all cases they have been careful to ensure that local engineers obtain the skills to enable Algeria to build satellites independently. For instance, in partnership with Surrey Satellite Technology Ltd., Algeria's AlSat-Nano was a "joint endeavour by the UK and Algeria" to build and operate a CubeSat, with the project "designed to provide training to Algerian students, making use of UK engineering and experience".[199] Also in partnership with Surrey, Algeria contracted the foreign manufacturer to "build the Alsat-1B Earth observation satellite in a transaction that includes substantial training and education of Algerians in satellite design".[200]

---

[196]See https://www.facebook.com/UWC-CPUT-Space-Association-442173742525308/.
[197]Danielle Wood and Annalisa Weigel, "Charting the Evolution of Satellite Programs in Developing Countries—The Space Technology Ladder," *Space Policy* 28 (2012): 17.
[198]Ibid.
[199]University of Surrey, *AlSat-1 N*, https://www.surrey.ac.uk/surrey-space-centre/missions/alsat-1n (accessed August 18, 2018).
[200]Peter B. de Selding, "SSTL To Build Alsat 1B Imaging Satellite in Algeria," *SpaceNews*, July 10, 2014, https://spacenews.com/41202sstl-to-build-alsat-1b-imaging-satellite-in-algeria/ (accessed August 19, 2018).

By following this approach, Algeria is building up a cadre of skilled engineers and technicians to make it increasingly independent in building and operating satellites.

Another example of space-related education furthering the SDGs is the two African Regional Centres for Space Science and Technology Education (ARCSSTE), one in English based in Nigeria, and one in French based in Morocco, established by UNOOSA. Having been established in 1998, the two African centres "run common curricula on four thematic areas of space science and technology education", including in areas such as satellite communication, basic space science and atmospheric physics, and remote sensing.[201] In the case of ARCSSTE-E in Nigeria, this centre has had positive knock-on effects that "either initiated or encouraged the establishment of space related courses in some tertiary institutions in Nigeria", and ARCSSTE-E "also led to the establishment of the Nigerian space agency called 'National Space Research and Development Agency (NASRDA)'".[202] The centre also undertakes outreach programmes aimed at the public, students, and policy makers, as well as workshops, seminars, and conferences. It further promotes awareness of, and knowledge in, international space law.[203] The ARCSSTE are examples of space-related education facilities becoming hubs for wider societal benefit, and since they "made giant strides in education and training of specialists" and in public and policy maker outreach, they support the broader SDGs through skills development and the promotion of the SD culture (including the role of science and technology in supporting that culture) discussed earlier.

Another example, within a tertiary education setting of space education supporting the SDGs is the partnership between South African universities related the Square Kilometre Array (SKA). The University of the Western Cape, the University of Cape Town, University of Pretoria, and Northwest University are collaborating through the Inter-University Institute for Data Intensive Astronomy (IDIA), that "responds to the big data challenge" of the SKA.[204] It is also an excellent demonstration of the many ways in which space-related education promotes skills across an array of fields, namely "astronomy, computer science, statistics and eResearch technologies" in order to "create data science capacity for leadership in the MeerKAT SKA precursor projects, other precursor and pathfinder programmes and SKA key science".[205] In other ways, the University of the Western Cape (UWC) is excelling in space sciences, with "an internationally-recognised research group which plays a leading role in aspects of MeerKAT and SKA science - and

---

[201]Olakunle Oladosu and Etim Offiong, "Improving Space Knowledge in Africa: The ARCSSTE-E," *Space Policy* 29 (2013): 154.

[202]Ibid., 155.

[203]Ibid., 156.

[204]University of the Western Cape, *Three SA Universities join forces to bolster SKA*, https://www.uwc.ac.za/News/Pages/Three-SA-universities-join-forces-to-bolster-SKA.aspx (accessed August 19, 2018); Nicklaus Kruger, "The Sky's No Limit: UWC Ranks In World Top 200 In Astronomy," *University of the Western Cape*, August 16, 2018, https://www.uwc.ac.za/News/Pages/The-Sky%E2%80%99s-No-Limit-UWC-Ranks-In-World-Top-200-In-Astronomy.aspx (accessed August 19, 2018).

[205]Ibid.

dozens of UWC researchers are involved in a variety of MeerKAT and SKA projects", recognised by the 2017/2018 URAP World University Rankings which placed UWC 162nd in the world in Astronomy & Astrophysics.[206] Advances in these broader fields of science, technology, and innovation are critical to the success of the SDGs as a whole, and it has been argued that they can contribute in a variety of ways including "[a]dvising on challenges", "[p]roviding indicators for monitoring progress, "[a]dvising on policies and actions", "[s]earching for innovative solutions", and "[e]nsuring every country and the UN have a robust science-policy interface".[207] Others have argued that without scientific and technological advances developing countries will be "guaranteed" to face continued poverty and depression, and that scientific illiteracy "is one of the world's great problems and an increasingly divisive factor between haves and have-nots" with "[m]uch of the quality of living increase and economic growth now depend[ing] on scientific and technical awareness and on an ability to incorporate new knowledge and devices into the economy and lives of individuals".[208]

A fourth example of space-related education producing skills to support the SDGs is university-based small satellite projects. South Africa's first satellite—SUNSAT (launched in 1999)—was built locally without foreign assistance, by graduate students of the Department of Electrical and Electronic Engineering at Stellenbosch University.[209] In this case, the effort led to the formation of a spin-off associate company of the university, SunSpace and Information Systems (Pty) Ltd in 2000, later absorbed into SpaceTeq, a division of the state-owned aerospace and defence conglomerate Denel.[210] SUNSAT serves as one example of "dozens of universities … now utilizing satellite projects as official educational tools".[211] This serves as a capacity building and human resource development effort, with wider societal implications in that

> Establishing a capacity in space technology development will not only contribute to improving the operational use of space applications, but will also enable countries to transition from being a passive space user to becoming a more active player in international space cooperation … and by actively contributing to the collection and sharing of space-based data and information in support of research activities and sustainable development.[212]

---

[206]Ibid.

[207]E. William Colglazier, "The Sustainable Development Goals: Roadmaps to Progress," *Science & Diplomacy*—American Association for the Advancement of Science, January 5, 2018, http://www.sciencediplomacy.org/editorial/2018/sdg-roadmaps (accessed August 19, 2018).

[208]Adriana Ocampo, Louis Friedman, and John Logsdon, "Why Space Science and Exploration Benefit Everyone," *Space Policy* 14 (1998): 138–139.

[209]European Space Agency (ESA), *SUNSAT (Stellenbosch University Satellite)*, 2002, https://earth.esa.int/web/eoportal/satellite-missions/s/sunsat (accessed August 19, 2018).

[210]Denel Spaceteq, "From SunSpace to Spaceteq," *Denel SOC Ltd.*, http://www.spaceteq.co.za/home/ (accessed August 19, 2018).

[211]John L. Polanksy and Mengu Cho, "A University-Based Model for Space-Related Capacity Building in Emerging Countries," *Space Policy* 36 (2016): 2.

[212]Werner Balogh, "Capacity Building in Space Technology Development: A New Initiative within the United Nations Programme on Space Applications," *Space Policy* 27 (2011): 182.

This is an excellent link to the third 'pillar' of the SEEMS model, namely how education (through SDG 4) supports the space sector.

4.3 SDG 4 and Education Supporting the African Space Sector

In recent years, African countries have been making "increasingly ambitious" efforts to expand their space science and technology because, as we have argued above, "[i]nvesting in space science and technology can serve as a key enabler for inclusive and sustainable development in Africa".[213] We have also echoed the argument of UNOOSA that space can motivate students to pursue studies within the sciences, and thus "[b]y reaching out to young people, building their interest in astronomy and other areas and enhancing scientific education in this way, we can bolster enrolment rates in Stem subjects".[214] Ultimately, the third 'pillar' of the SEEMS model focuses attention on the way in which these educational advances can then

> help build the critical mass of scientists, researchers and engineers on whom the success of the AU's space strategy depends. At the same time … equipping the next generation with high-level scientific and technical competencies will help African nations train and retain the highly skilled personnel necessary to enhance productive sectors and boost our continent's structural economic transformation.[215]

This neatly illustrates the symbiotic relationship between space and education in relation to SDG 4 in Africa. Space science, technology, data, and activities support education and hold the great promise that, as the South African Freedom Charter proclaimed in 1955, "The Doors of Learning and Culture Shall be Opened!"[216]— including doors of learning in space-related fields, which then in turn produce the skills necessary to both support SD and generate the next wave of advances in space that then feed back into supporting education and motivating the next round of students. This is especially critical, as Abiodun argues, since there has historically been a general attitude of neglect of African states vis-à-vis science and technology, and "[c]ompliance or lack of it to the allocation of 1% of GDP in each African country to science and technology, as agreed to in the Lagos Plan of Action, is a case in point".[217] There has been, and continues to be, a dire need that "Africa must prioritise the education and training of the scientists, technicians, engineers and mathematicians it needs for both national and collective regional space efforts and

---

[213]Paul Boateng, "Outer Space is the Place for Africa's Future," *Mail & Guardian Online*, May 26, 2016, 1–2, https://mg.co.za/article/2016-05-26-00-outer-space-is-the-place-for-africas-future (accessed August 19, 2018).
[214]Ibid., 3.
[215]Ibid., 3.
[216]South African History Online, *The Freedom Charter*, August 4, 2016, http://www.sahistory.org.za/article/freedom-charter (accessed August 19, 2018).
[217]Adigun Ade Abiodun, "Trends in the Global Space Arena—Impact on Africa and Africa's Response," *Space Policy* 28 (2012): 288.

related social and economic development activities".[218] The case of Algeria can again serve as example here. In 2006 a deal was struck between Algerian authorities and European firm EADS Astrium, whereby two satellites would be built, and the first (AlSat-2A) would be integrated and tested in France, with the "provision for Algerian engineers to work side-by-side with the EADS Astrium development team, with intensive training given in space technology", and based on this training the second satellite (AlSat-2B) would be integrated at an Algerian facility.[219] This is a clear example of space-related education efforts fostering the development of the African space sector. Accordingly, we argue that by supporting education in all the ways discussed earlier, the space sector can thus also help to meet its own need for scientists, technicians, engineers, and mathematicians, thus advancing its own cause through the SDGs, and bringing the African education-space ecosystem full circle. The next section will consider some of the ways in which the future advancement of SDG 4 via space can be further promoted, in light of the discussion throughout this chapter.

5   Future Advancement of SDG 4 Via Space

This section will present a number of recommendations for further leveraging the space sector to promote African tertiary education in line with SDG 4. First, using the example of e-learning, space itself can be used as a learning tool to promote STEM skill development. With the massification of higher education in Africa, and the use of ODeL to help meet the needs related to this, the use of a variety of educational tools related to space can be expanded to meet the needs of the SDGs related to relevant skills, including technical and vocational skills and ICTs. For ESD+ to be successful, the African continent needs constructive alignment between up-to-date technology (hardware), subject matter experts (SMEs) within the educational field, up-to-date educational technology for the learning and teaching environments, and educational technologists who can work with SMEs for the implementation of technology within educational curricula. Furthermore, in terms of ESD+, the role of formal and non-formal education and training for youth and adults searching for a career in space sciences, or the sciences in general, is an underdeveloped area, and thus better infrastructure (Internet, etc.) should help the training and further education along. Specific tools that would be useful in this educational context, which are directly related to the space arena, include Stellarium Astronomy Software, a free open source planetarium offering a realistic 3D sky view with a "default catalogue of over 600,000 stars" and "extra catalogues with more than 177 million stars", a "default catalogue of over 80,000 deep-sky objects" and an "extra catalogue with more than 1 million deep-sky objects", and offering

---

[218]Ibid., 289.
[219]European Space Agency (ESA), AlSat-2 (Algeria Satellite-2), https://directory.eoportal.org/web/eoportal/satellite-missions/content/-/article/alsat-2 (accessed August 19, 2018).

constellations for over 20 different cultures.[220] Another example is Celestia Software, which is a free space simulation that "lets you explore our universe in three dimensions", offering an interactive environment ranging from "planets and moons to star clusters and galaxies, [where] you can visit every object in the expandable database and view it from any point in space and time".[221] Free educational tools such as these should be increasingly leveraged to expand access to, and understanding of, space knowledge as more people become digitally connected in Africa. Another way in which space can further support the SDGs in terms of education is MOOCs, such as those offered by edX, which was founded by Harvard University and MIT in 2012.[222] Space-related courses such as *Introduction to Aerospace Engineering: Astronautics and Human Spaceflight* can be audited for free, and it is strongly recommended for the future African E-University to include such courses in languages accessible to the majority of Africans. It will also be vitally important to leverage other tools such as JAWS Screen Reader, which is a powerful tool for those "whose vision loss prevents them from seeing screen content or navigating with a mouse" and "provides speech and Braille output for the most popular computer applications",[223] and others like Dragon speech recognition software[224] to promote equity in educational access.

A second recommendation echoes the call of Newport that universities and other institutions of higher learning should look to satellite communications in order to expand space-related education for students.[225] It is correctly argued that "it would be impractical for the majority of institutions to design and build the space portion (i.e. the satellite) because of lack of time, expertise and funds", but that other viable and affordable options include communicating with satellites already in orbit, such as those of the Amateur Satellite Service,[226] since

> The equipment used in contacting amateur satellites is considerably cheaper than those used in commercial environments. When redundancy and extreme reliability are not required, equipment used in ground segments of such stations need to be no more expensive than what is affordable with a university laboratory budget. ... Simple satellite ground stations can be constructed for several hundred dollars and still serve as an adequate platform for educating technicians and engineers.[227]

---

[220]https://stellarium.org/ (accessed August 19, 2018).
[221]https://celestia.space/ (accessed August 19, 2018).
[222]edX Inc., *Our Story*, 2018, https://www.edx.org/about-us (accessed August 19, 2018).
[223]https://www.freedomscientific.com/Products/Blindness/JAWS (accessed August 19, 2018).
[224]https://www.nuance.com/dragon.html (accessed August 19, 2018).
[225]Jonathan Newport, "Amateur satellites: A neglected vehicle for satellite communication education," *Space Policy* 21 (2005): 101–104.
[226]For example the South African Radio Satellite Association, http://www.amsatsa.org.za/ which aims to "promote the use of amateur satellites and to encourage active experimentation in satellite communication and allied field of experimentation".
[227]Newport, "Amateur satellites," 102.

Newport argues that this "underutilized amateur satellite service provides a framework for this investment and is begging to be exploited for global education", and can produce "many tangible and intangible rewards for individuals, institutions and countries[228]

> A vast range of equipment is available for use in contacting amateur satellites. The spectrum covers low-power handheld radios transmitting and receiving through manually controlled beam antennae to commercial-grade communication transceivers pumping hundreds of watts into automated high-gain satellite dishes. It is a misconception that communication with orbiting satellites is expensive. One needs only meager resources and ingenuity to perform basic satellite communication and not a great deal more to work with more advanced satellite functions.[229]

The point here is that it remains important to combat the misconception and criticism that is still prevalent in Africa, whereby it is "questioned why countries that are recipients of overseas aid are investing in expensive space adventures",[230] since "[a]ssuming that the subjects of modern science are not relevant to a poor or illiterate population only assures they will stay that way".[231]

A third recommendation is for academics within HEIs to collaboratively work with secondary school teachers to showcase first-year university-level physics experiments, worksheets, and activities, and to give the students at secondary schools a sound foundation of what is expected of them at tertiary level. After all, success at tertiary level heavily depends on the foundational skills established in earlier education. Further teacher training is also a critical component of this, and by using a combination of face-to-face and online teacher training courses it is possible to meet the need highlighted in the Incheon Declaration, to "[m]ake learning spaces and environments for non-formal and adult learning and education widely available, including networks of community learning centres and spaces and provision for access to ICT resources as essential elements of lifelong learning".[232] There also continues to be a need within higher education for e-learning 'champions' to collaborate with their colleagues to empower learning and teaching.

Further community outreach will also be important to overcome misconceptions about space studies in rural or traditional communities, or where the feeling may persist that the developments of SD are far removed from daily reality. Overcoming this barrier for space sciences will not be an easy task, but with continuous efforts, positive strides can be taken, such as (i) including rural communities in the decisions related to SD programs in Africa, (ii) space science centres in Africa collaborating with international partners for scholarships on African-related space science questions, and (iii) giving African space science research a priority at

---

[228]Ibid., 104.

[229]Ibid., 103.

[230]Boateng "Outer Space is the Place for Africa's Future," 2.

[231]Ocampo, Friedman, and Logsdon, "Why Space Science and Exploration Benefit Everyone," 138.

[232]World Education Forum, *Incheon Declaration and Framework for Action for the implementation of Sustainable Development Goal*, 52.

international conferences and symposia, which will aid these African researchers to develop themselves further and raise the profile of African space science.

Fourth, the establishment of the fifth institute of the PAU relating to space sciences should be kept on track and that it should make space studies as accessible to students from across the continent as possible, even if they cannot attend face-to-face lectures. This institute should thus take cognizance of the arguments presented throughout this chapter and ensure that the 'doors of learning shall be opened' in a way that maximises accessibility to all Africans and ensures quality education practices both in and for SD, as we have argued.

Finally, as Internet connectivity continues to increase in Africa, especially if the satellite constellations prove successful, basic information skills will be in urgent demand. This is a place where both education and space sciences can benefit. As Edejer argues,

> even if the woman in the village has access to the Internet, she will not necessarily be able to use the information to improve her child's health [or education] because trying to get information from the internet is like drinking from a firehose – you don't even know what the source of the water is.[233]

Therefore, even if there is Internet, effectively using it is another matter, and thus beyond access, the ability to meaningfully interpret and evaluate information, and to contextualise it—in other words information and knowledge management skills —will also be fundamental to Africa's success in terms of SDG 4. Tapping into, and making use of, existing skills on the continent is therefore vitally important. As Zozie argues by way of an example, "[d]igital universities can succeed if universities in Africa can form networks for education and research".[234] Ultimately, what is needed is to continue to raise awareness within Africa of the many symbiotic and interrelated ways in which space and education can support each other for SD, making continued research in this field important.

6   Conclusion

This chapter has sought to illustrate how interdependent space and the educational SDGs are in Africa by way of the tertiary education sector, and by making special reference to e-learning as one of the ways in which space technology can facilitate educational delivery to meet the needs of massification on the continent. It was also argued that the relationship between space and education constitutes a symbiotic ecosystem that also ultimately supports the growth and success of the space sector in Africa. As we argued earlier, one of the learning outcomes related to SD is systemic thinking, which is also necessary when analysing the ways in which space interacts with, and supports, the SDGs. It was also argued that the SDGs do not

---

[233]Tessa Tan-Torres Edejer, "Disseminating health information in developing countries: the role of the internet," *BMJ* 321 (2000): 798.
[234]Zozie and Chawinga, "Mapping an open digital university in Malawi," 223.

stand in isolation, but also rely on regional and national development strategies for their implementation.

A variety of models were presented to support our arguments in relation to what 'quality' education means in general (the education quality equation) and specifically within e-learning that represents an operationalisation of ESD+ (sustainable e-learning). It was thus emphasised that beyond traditional views of ESD, which in reality only look at the learning outcomes and learning content to promote skills for SD, it is also necessary to look at the means of learning itself to place education on a sustainable trajectory to meet the urgent and serious challenges facing Africa. The final model focused on the symbiotic interrelationship between space and SDG 4, but as we argued, this also serves the ends of Agenda 2063 and national development strategies. Space is thus not only a means for enacting the SDGs, but an end in itself to generate skills in short supply and to advance African space ambitions.

Africa faces many challenges if it is to avoid repeating the experience of the MDGs where unfulfilled goals marred an otherwise noble endeavour to end the worst miseries of poverty and to empower communities to take their destinies into their own hands. The role of space in supporting the SDGs needs far more recognition if the 2030 targets are to be met, and if Africa's continental vision and national development strategies are to be successful. Space can no longer be viewed as an optional factor or luxury in the SD equation, and if the African space sector itself is to change lingering misconceptions about its value for African societies, or find the skills it will need to take its rightful place on the global stage, it is incumbent on the sector and its professionals to continue to seek out ways in which to support education and educational technologies on the continent. We as Africans need both the science mentioned by Pasteur, and the critical faculties mentioned by Sagan, to make our dreams for our future a reality.

Author Biographies

André Siebrits has a Master of Arts in International Studies from the University of Stellenbosch, and is currently a Ph.D. Candidate at the Department of Political Studies at the University of Cape Town. He has experience as an e-learning researcher and as an African political risk analyst. He is currently conducting research for the European Space Policy Institute (ESPI).

Valentino van de Heyde has a Master of Science in physics (Space Weather Physics), and is an academic developer and Ph.D. Candidate at the Department of Physics and Astronomy at the University of the Western Cape. He has lectured undergraduate courses within the department, and has experience as an Instructional Designer.

The Possible Beneficial Effect of Using Small Satellite Technology to Promote the Achievement of the UN Sustainable Development Goal of Poverty Reduction Specifically on the African Continent

Anton de Waal Alberts

Abstract

The United Nations Sustainable Development Goals are hard to achieve, especially so for African states. However, the African continent is brimming with potential given its arable land, vast mineral deposits, and demographic dividend in the form of a generally young population. The current developmental status of the continent, while lagging in comparison with many other developing continents and regions, presents an opportunity to implement the most advanced and contemporary technologies for developmental purposes due to the lack of legacy technologies that need to be replaced. This study explores the proposition by investigating the possible achievement of the United Nations Sustainable Goal 1, namely the eradication of poverty, by making use of small satellites within swarm configurations. The investigation is performed by making use of systems theory and reframes the question by investigating the possible effect that the techno-sphere dimension of small satellite technology could possibly have on the African continent's poverty problem given the fact that the continent and its constituent parts form a complex system.

A. de WaalAlberts (✉)
University of Johannesburg, Johannesburg, South Africa
e-mail: anton.alberts@prevoyance.co.za

A. de WaalAlberts
University of Stellenbosch, Stellenbosch, South Africa

© Springer Nature Switzerland AG 2019
A. Froehlich (ed.), *Embedding Space in African Society*, Southern Space Studies,
https://doi.org/10.1007/978-3-030-06040-4_11

1   Introduction

The African continent and its states represent some of the poorest areas in the world. The United Nations Sustainable Development Goals (SDGs) are, given its developmental nature, especially relevant to Africa. This is evidenced, amongst others, by the reference under the heading of Sustainable Goal 1 to poverty in general, but also "in particular sub-Saharan Africa".[1]

Economic development is a prerequisite for poverty eradication. Economic development that democratises wealth among the broad population can result in poverty reduction but should directly and indirectly also address the other 16 SDGs as they all relate to each other.[2] This implies that the SDGs are evidence of emergent properties of the world from which they arise and indicates the complex nature of the African continent and its constituent parts.

Despite its developmental challenges, the African continent is brimming with potential factors that contribute to unlocking economic value and eradicating of poverty, like arable land for agriculture, vast mineral deposits, and a young population that can provide the labour needed for economic development. These factors, amongst others, need to be included in any economic planning.

One potential factor that could possibly accelerate economic development on the African continent is ironically emergent from the fact that the continent is generally less developed than the rest of the world and due to its lag in technological adoption —as opposed to the early adopters since the Industrial Revolution—can frog-leap the rest of the world technologically. In this respect satellite technology can play a role, especially the most recent developments that allow for the utilisation of more features at less cost in the form of small satellite systems.

This study, therefore, addresses the possible leverage or influence that small satellite technology can have on the economic development of the African continent and its constituent parts. This will be performed by using systems theory as a framework of analysis.

2   The World as a Complex System

The world is complex with many interlinking branches that creates feedback cycles that add even more layers to the complexity. One way of understanding this is by looking at the world through the frame of systems theory.

---

[1]United Nations Department of Economic and Social Affairs, *United Nations Sustainable Development Goals*, undated, https://sustainabledevelopment.un.org/sdg1 (accessed on 20 July 2018).
[2]United Nations Department of Economic and Social Affairs, *United Nations Sustainable Development Knowledge Platform*, undated, https://sustainabledevelopment.un.org/ (accessed on 20 July 2018).

Systems theory provides a description of the nature of the universe as it is and how it is organised.[3] In general it allows for the identification of regularities and the formulation of rules or laws based thereon. These rules allow the modelling of complex interactions arising from diverse domains from the natural to the social world. Many types of systems theories exist with differing perspectives on the regularities. In this case the Biomatrix theory will be used to provide a platform on which to base the investigation herein. Biomatrix theory "integrates the key concepts of the whole field of systems thinking into one coherent theory". Only the most relevant aspects are referenced and used in this study.

The Biomatrix theory divides the world into sub-webs. These three interacting sub-webs are hierarchically organised and are constituted by the nature-sphere (i.e. nature's interacting systems from the universal/galaxy/solar/planetary to the sub-atomic level), the psycho-socio-sphere (i.e. psychological and social systems that arises from the nature-sphere at the level of individual human and animal mental activity and social interaction, e.g. political systems, economic systems, cultural systems etc.), and the techno-sphere (i.e. technological systems that emerges from the psycho-socio-sphere comprising all the artefacts produced by humans and animals).[4]

The African continent consists of these interacting sub-webs—as the remainder of planet Earth. However, not only does the continent constitute a system, but as it is an interacting system of systems, it is also a complex system. Complex systems are difficult to understand, manage and influence. Due to the characteristic of non-linearity, lack of equilibrium and feedback loops, amongst others, these systems tend to be opaque and small changes in initial conditions can lead to unexpected large causal effects.[5]

Another important aspect of the Biomatrix is that it uses the concept of "mei" which is a description of the fundamental and inseparable fields of the cosmos, namely matter, energy and information. These fields are, therefore, to be found within all systems and will be alluded to in the discussion below.

In reflecting on the question posed in this study, cognisance must be taken of the complex nature of the African continent and its parts in the form of nation states, as well as larger and smaller regions. This allows the question to be reframed as follows: Is it possible that a techno-sphere artefact like small satellite systems can be a factor to improve the socio-economic conditions of the human psycho-socio-sphere in the form of the reduction of poverty?

---

[3]Dostal et al., 2005. Biomatrix. A Systems Approach to Organisational and Societal Change. Cape Town: Mega Digital.
[4]See Footnote 3.
[5]See Footnote 3.

3 Satellite Systems

3.1 Brief History and Nature

The development and launch of satellites have historically been the sole domain of space-faring states, like the USSR (today Russia), the USA, Europe, Japan, India and China. The services provided by these satellites in the form of communication, broadcasting and remote sensing was over time opened to non-space-faring states. From the 1980s onwards an informal amateur satellite community developed in the form of, amongst others, universities and technological institutes. It was they who started to develop affordable small satellites for research and experimental use as opposed to the huge satellites used by the global state- and commercial actors.[6]

Small satellites were built making use of self-made and off-the-shelf technologies and did at first not emphasise rigorous engineering and testing to reduce the risk of failure as is the case with large satellites. However, over time small satellites were developed to provide commercial applications and services in competition with larger satellites. Today small satellites are becoming smaller still with more applications and uses that can be accessed on the global market. Today small satellites are considered a class of its own with even smaller satellites being developed, known as pico, nano and micro satellites (described *infra*).

Wood and Weigel outlines the nature of the small satellite industry as follows[7]:

- Small satellites are built to have lower performance expectations, greater risk of failure, and is able—due to its shorter lifecycles—to use newer technologies at a faster rate, like cell phone technology. The shorter lifecycles are a feature of the use of technologies designed to be used on earth and not the extreme environment of outer space;
- Small satellites tend to have only one payload;
- Missions are designed to be shorter;
- Small satellites are created by flat organisations in comparison with the large bureaucracies that develop large satellites;
- Small satellites are designed to carry fewer instruments.

Satellites, in general, as systems (a combination of interconnected materials, energy and information) are designed to provide services that (i) communicates information (ii) broadcasts information, and (iii) collects information about material and energy of the earth and its surrounding environment. Information in general is of great value for human social systems where it is used to plan and create strategies. Satellite systems, therefore, contribute to the creation of social systems like agriculture, cities, and infrastructure generally on the one hand, and the

---

[6]Wood D, Weigel A. 2014. Architecture of small satellite programs in developing countries. *Acta Astronautica*. Elsevier Ltd.
[7]See Footnote 6.

management of events, like epidemics, natural disasters, wars, displacement of humans and animals, and action towards climate change on the other hand.

The quality of the services provided by satellites—small and large—are ever increasing and new abilities should develop over time. An overview of services provided now and in the future by all satellite-types is set out below.

3.2 Types and Uses of Satellites

The current typology and uses are based on the analysis of Wood and Weigel.[8]

3.2.1 Current Types and Uses

Satellites currently consist of the following types with concomitant uses:

- **Traditional Commercial and Government Satellites**: greater than 1000 kg with services relating to high resolution earth imaging, scientific measurement, communication, positioning and timing, reconnaissance and surveillance, with a lifespan often greater than ten years. These satellites can operate in the geostationary orbit.
- **Small Satellites**: approximately 100–1000 kg with services relating to earth imaging and scientific measurement in lower orbit, with a lifespan of three to seven years.
- **Pico, Nano and Micro Satellites**: less than 100 kg with services relating to technology demonstration, mission science, non-real time communication and education in low-earth orbit, with a lifespan often less than one year. Micro-satellites are considered to be between 10 and 100 kg, nano-satellites between 1 and 10 kg, and pico-satellites less than 1 kg. A form of nano-satellite is known as a CubeSat that conforms to a specific size of ten cubic centimetres with a mass of about 1 kg.

3.2.2 Future Uses

The technological capability of satellites is in continuous development. Currently small satellites, due to their quick adoption of new technologies, are showing the most evidence of expanding capabilities. While it is usually impossible for small satellites to perform all of the services provided by large satellites, the new trend of developing swarms or constellations of small satellites (a multi-satellite system) can result in them also providing the services of large satellites, such as real-time communication, broadcasting, and positioning and timing, amongst others.

---

[8]See Footnote 6.

4   African Continent Satellite Technology Development and Use

The African continent's development of satellite technology is lagging that of most other continents and regions. Like the rest of the world, it does however make use of the services provided by large satellites for communication, broadcasting and remote-sensing. In terms of small satellite development and use, the continent is still in the process of development and adoption of use of the services, as much as the rest of the world. Interestingly, Africa is actively participating in the unfolding development of small satellites themselves as will be elucidated below.

The African continent's development of its own satellite capacity will now be discussed.

4.1   Satellite Development in Africa

Very few African states—as the social constituent subunits of the continent—have a space program or is involved in satellite development on its own or in partnership with other states or institutions. However, as these activities are normally linked to a state's socio-economic development level, the fact that there are serious satellite projects taking place in various states, serves as a testimony to the African continent's strong will to participate in fields of high-technology.

4.1.1   Large and Small Satellite Development in Africa

Several African states have engaged in the building of their own satellite capacity by either procuring satellites from other states or companies, or by developing their own with its own or shared resources. Intra-African partnerships are also planned or in process.

The following states have engaged in the development of satellite capacity:

- **Algeria**: The country has six satellites in orbit. All of them have been procured from spacefaring-states and their internal private product providers. All the satellites are remote-sensing in nature, save for the last one launched in 2016 that is a geostationary communications satellite. All satellites can be considered large satellites, save for one that is a CubeSat built as part of a joint educational mission between the Algerian and UK space agencies.[9]
- **Egypt**: The country has an earth observation satellite in orbit procured by the government from the Ukraine and three communication satellites procured by a quasi-commercial entity (the NileSat organisation) from EADS Astrium in Europe. All satellites are large ones.[10]

---

[9]Algerian Space Agency, *Algerian Satellites*, 2017, http://www.asal.dz/ (accessed on 13 August 2013).
[10]NileSat, *The Satellites*, 2018, http://www.nilesat.com.eg/ (accessed on 13 August 2018).

- **Kenya**: The country launched its first satellite in 2018. It is a CubeSat developed by the University of Nairobi and launched from the Japanese module of the International Space Station.[11]
- **Nigeria**: The country has three earth observation satellites and one communications satellite. All satellites can be considered as large satellites.[12]
- **South Africa**: The country's first satellite, SUNSAT, was initiated and developed by the University of Stellenbosch and launched in 1999. It was used for remote sensing with a focus on educational capacity-building, experimentation and satellite product development capacity. Its second satellite, SumbandilaSat, was a joint-venture between a private sector player (as a spin-off from the SUNSAT project) and the government. This satellite was developed for more advanced remote-sensing and was launched in 2009.[13] Currently the Cape Peninsula University is developing a CubeSat. All the satellites are small satellites.

It is interesting to note the following two trends from the above information:

- Most satellites launched are large satellites procured from spacefaring developed states;
- Later satellites include small satellites, like CubeSats, but they are still outnumbered by the large satellites.

5 Africa and SDG 1

To address SDG 1, namely eradication of poverty, the economic system must be aligned to this goal. The economic system of Africa is nested in the larger political system of the African continent as emergent from the political system of each African state. This means that the economic system of the African continent is emergent from the economic system of each African state. The alignment of the economic system so that it can grow, create value and employment opportunities, and address the poverty problem, is dependent on information whereby the levers of energy can be used to direct matter, energy and information for the creation of a conducive economic environment. In this regard, satellite technology can play a decisive role.

---

[11]Winnick E, 2018, Kenya's first satellite is now in Earth orbit, *MIT Technology Review*, https://www.technologyreview.com/the-download/611127/kenyas-first-satellite-is-now-in-earth-orbit/ (accessed on 13 August 2018).

[12]Ministry of Science and Technology, *Nigeria National Space Research and Development Agency*, 2018, http://nasrda.gov.ng/en/missions/ (accessed on 13 August 2018).

[13]South African National Space Agency, *South African Satellites*, 2015, http://atlas.sansa.org.za/atlas-sa_satellites.html (accessed on 13 August 2018).

5.1   Satellites, Information and Economic Development

The economic system of each African state and of the African continent relies on information to function effectively and for political, economic and financial institutions to make decisions that influence the status of the economic system. In practice that information is available, amongst others, in the form of statistics about population sizes, population wealth and spending patterns, industry status and development, various indexes and trade numbers.

Satellite-sourced information can enhance the above information, but also add new types of information that can increase the ability to plan and make strategic decisions. High-resolution earth imaging and scientific measuring can provide information, amongst others, about the African continent's agricultural spaces, city development, natural systems like water sources, climate, human and animal population movement, and monitor natural systems for early disaster-warning. This information can be used in isolation or in conjunction with general existing statistics. The information can be plotted onto maps as part of a geographic information system that can increase the ease of use. In this respect African states and the continent can become part of the developing Big Data economy.

(This interaction between satellites, its generated information and use in the fields of social and natural systems, serves as an apt example of the interaction between the Biomatrix spheres of the nature-sphere, psycho-social-sphere and the techno-sphere.)

Over and above planning by making use of satellite-sourced information, African states and the continent will add another dimension to their economic development dynamics by investing in their own satellite technology and ability. Owning the capability to perform various satellite functions will not only provide instant access to crucial information but will also make African states and the continent a space player that can compete in the global market to sell satellite services and products. This will increase economic activity in the respective economic systems of the satellite-empowered African state[s] and indirectly the continent itself. This economic activity can be characterised as twofold:

- **Direct economic activity:** Selling of satellite services and products; and
- **Indirect economic activity:** Using the data alone or in conjunction with other data (Big Data) to develop insight for strategic decisions for itself; selling satellite-generated data alone or in conjunction with other data (participating in the Big Data economy); creating new economic spaces in the broadband internet sector with developments like the internet-of-things, online streaming services and blockchain services; and creating further economic spaces in the transfer of data/information and digital broadcasting sectors.

5.2 Satellites for Africa

Traditionally, the cost and limited technological ability were inhibiting factors to the development of satellite capacity for African states and the continent, but the small satellite revolution has changed that. As small satellite technology develops it will soon be possible to provide services that large satellites provide, like communications for broadband internet connection amongst others, broadcasting, reconnaissance, and real-time positioning and timing. This will be achieved by parking swarms of small satellites in orbit that forms a network of functionality and purpose. The technology for small satellites is more readily available, cost-effective and more advanced than those found in traditional large satellites. Unrelated technological developments like cell phone hardware and software can have a positive effect on satellite development as these technologies can be adapted for use in small satellites.

The question then is whether African states and the continent can become satellite system owners as opposed to users of expensive systems owned by First World states? In order to answer this question, it will be apt to have regard to the historical models of satellite system creation used by African states. Wood and Weigel have performed extensive research on the architectures of small satellite programs in developing countries.[14]

Their research was focused on collaborative satellite development projects (CSDPs) within developing countries, including Africa. Their research concluded that three archetypal models of CSDP can be discerned and that context plays a significant role in differentiating between the CSDPs with reference to project initiation, supplier selection, technology characteristics and training approaches.

The three archetypal models are as follows.

5.2.1 The Politically Pushed Project

In this project strong political will from a key national leader exists. The architecture of the project exhibits the following consistent aspects:

- Low effort to initiate and approve the project due to the existing political will;
- Focus and effort is on defining the project activities and technology, like choosing the foreign supplier with a common vision as the government;
- A conservative technological approach as the satellite need not be highly ambitious, but merely successful to introduce the state's first national project to the world;
- The foreign supplier guides the local space organisation due to their inexperience on satellite development;
- Informal mentorship is provided by the supplier's engineers to the procuring implementor's engineers. In the main the procurer's engineers are mostly learning skills as opposed to a contributing to the project.

---

[14]See Footnote 6.

The following African states followed this model of satellite development: Algeria, Egypt, Nigeria.

5.2.2 The Structured Project

This project may consist of the follow-up project after the Politically Pushed Project. The systems architecture of this project exhibits the following aspects:

- This time there is low political support. This time a leader from the local space organisation must drive the project and convince politicians that it is economically and otherwise palatable;
- The foreign supplier is chosen through a more objective and rigorous process as opposed to the politically pushed project that based its choice of supplier on a more personal basis;
- The satellite project is highly technical with high performance parameters;
- The local engineers are more involved in the various development phases of the satellite and skills transfer is regimented and tracked to ensure successful upskilling.

The following are examples of African states that have engaged in this model of satellite development: Algeria, Egypt, Nigeria.

5.2.3 The Risk-Taking Project

This project form is also usually used after initial satellite projects have been engaged in. However, there are instances where this project model was used as the very first satellite project. The systems architecture exhibits the following aspects:

- The political will is this time round at best mediocre. A leader from the local space organisation usually takes the initiative to demonstrate the need for the project in an innovative way to ensure government approval;
- If a foreign supplier is chosen, it is usually performed based on trust as a technical risk is taken in this case;
- The technical nature of the project is normally high risk for high reward;
- Mentorship may be formal or informal with practical and on-the-job training.

The following are examples of African states that have engaged in this model of satellite development: South Africa, Algeria and Kenya.

States are seen to transition from the first model to the two others. However, there are two more models that deserve mention.

5.2.4 The Higher Education Project

This project is one that can develop without any government support or in partnership with government. It can also precede or follow any national government project. The purpose of these projects usually is educational and researched-based, but nothing prohibits a higher educational institution to develop a satellite for commercial applications.

The following African states engaged in this type of project: South Africa and Kenya.

5.2.5 The Regional Project

This project is based on partnerships between states whereby all contribute to the development of a satellite project which can consist of one or more satellites. The regional partnership can also develop the project on its own or procure components, services and assistance from established satellite developers. A region does not necessarily have to consist of states adjacent to each other, but also geographically dispersed states on the African continent.

A good example of this in Africa is the envisaged African Resource Management Satellite (ARMS) Constellation that arose from "a need for regular high resolution data over Africa for resource management operations".[15] ARMS consist of an inter-state partnership between various African states to build a constellation of satellites. Those states are South Africa, Nigeria, Algeria, Egypt and Kenya. Unfortunately, ARMS has not really made any progress since its inception.

5.3 African Satellite Development Systems

The above exposition is historical in nature and provides guidance on what satellite development systems have been operationalised in Africa. There are, however, also forms of projects that have not yet taken place in Africa but are witnessed in or among other spacefaring states. They can be listed as follows (with examples):

- **Intra-Governmental (within the state)**:

  - **Government-Driven Public-Private Partnerships**: National Aeronautics and Space Administration of the USA in partnership with Boeing and Lockheed Martin to develop the new manned rocket for the planned Moon and Mars missions;
  - **Business-Driven Public-Private Partnerships**: SpaceX's rocket and supply vehicle assets in partnership with NASA to supply the International Space Station;
  - **Business-Driven**: Virgin Galactic's space tourism project and Blue Origin's manned spaceship project, amongst others.

- **Inter-Government (between states)**:

  - **Government-Driven Public-Private Partnerships**: European Space Agency
  - **Business-Driven Public-Private Partnerships**.

---

[15]Mostert S, 2008. The African Resource Management (ARM) Satellite Constellation. *African Skies*.

Given the above African and extra-African satellite development systems exposition, the following three overarching satellite system development configurations for use in Africa can be discerned:

- **Government-Driven**: includes all CSDPs, the Regional Project, and the Higher Education Project Partnership;
- **Non-Government-Driven**: Higher Education Project and Business Driven Projects;
- **Public-Private Partnerships**: Government-Driven or Business-Driven.

Any of these three overarching development systems can be used for satellite project development in individual African states or regional- or continent-wide. In effect individual African states can opt to develop their own constellation of small satellites as government alone or in partnership with other institutions and/or private business. African states can also opt to develop projects together with other African states in a regional project or continent-wide. Cooperation can also take place between African states and their private businesses and/or higher educational institutions. In general, due to the number of agents already in the field, a system of cooperation can emerge over time with hugely beneficial properties.

While African states can strive towards individual satellite capability development, it is more ideal to cooperate across regions or the continent as a whole due to SDG 1's continent-wide prevalence and the effect of sharing costs towards the project.

6 A Proposed Way Forward

While various African states have engaged in their own satellite projects by using any of the three overarching satellite system development configurations, none have engaged in acquiring or building small swarm or constellation satellite systems. As stated *supra*, these systems can position African states to sell services and products currently purveyed by developed states via large satellite systems and will provide African access to Big Data for planning and strategic decision-making.

Given the importance of global communications and other satellite services and the economic spaces and opportunities it creates, it is possible that owning and operating a small swarm satellite system in part, or as a whole, should inject economic energy into the African economic system. Such a small swarm satellite system can thus play a role in addressing the poverty challenge outlined in SDG 1 and indirectly the other sixteen SDG challenges due to their systemic connection to SDG 1.

Such a satellite system—depending on its orbital positioning - should be able to provide global or near-global products and services including much-needed data and information for the African continent itself. The initial costs should be absorbed over time as the small satellite system starts to deliver products and services to clients.

The important question, however, is how to get to the developmental point of a small swarm satellite system? The answer is to be found in any of the three overarching satellite system development configurations. Any of these configurations can lead to the proposed development and the choice will differ from state to state. In most cases the development will be Government-Driven configuration due to the developmental status of the various African state-economic systems. Little private enterprise in the space sector exists, save for perhaps South Africa.

The scope and size of the small swarm satellite system will also dictate whether one state will drive the project or whether a regional or other block of African inter-state cooperation will be necessary. Where one African state drives the project, it could follow the path set out by the High-Risk Project CSDP. Should the scope and size (and thus cost) be large, then the path of the Regional Project could be taken. While states would usually drive projects of this magnitude, there is nothing inhibiting the higher education institutions to engage in a Higher Education Project with commercial uses or a Business-Driven Project to arise. Historical data from South Africa does indicate that the odds are better for the afore mentioned to take place than the last mentioned.

Having said this, the ideal African small satellite project with the highest probability to influence the African economic system to eradicate poverty as per SDG 1, should take on the following form:

- **Objective**: the scope and size of the project must be to create a small satellite swarm system that is large enough to cover the whole or a significant part of the earth with its products and services, including the whole African continent;
- **Effect**: the larger the scope and size of the project, the greater the economic effect it could have on the African states involved and the African continent;
- **Satellite Development System**: given the large size, scope, cost, and the possible effect on the various economic systems across the African continent, one state should not take on such a project on its own. It should also be Government-Driven due to its economic importance. Therefore, it is proposed it should be a Regional Project consisting of a network of Public-Private Partnerships.

It is admittedly an ambitious project and will require strong political will to initiate and drive to its conclusion but can bring Africa closer to SDG 1. It is submitted that it is worth further investigation given the possible economic benefits that could emerge.

7 Conclusion

Africa, especially sub-Saharan Africa, is facing a huge challenge in meeting SDG 1. While it requires a multipronged approach given the complexity of the problem, its lag in development can allow it to leapfrog in technological application that may assist in economic development. One such technological system is satellite technology that can create data and information to be used in strategic decision-making, create revenues by way of selling products and services and can create economic spaces for new business to flourish. The most cost-effective satellite system that can create the same value and more than the traditional large satellite systems, is that of small satellite systems in the form of small satellite swarm systems consisting of a constellation of small satellites.

Various satellite development systems have been used globally over time. This study identified three overarching satellite system development configurations available for use by African states and discussed the various models that can be used.

A suggestion is made regarding the satellite system development configuration and model to be used to establish the highest probability to influence the African economic system to eradicate poverty as per SDG 1. It is proposed that the Government-Driven configuration in the form of the Regional Project model be used to create a small satellite swarm system that is large enough to cover the whole or a significant part of the earth with its products and services, including the whole African continent. This objective is to create a techno-spherical system that can have the most profound positive effect on the various economic systems on the African continent and the emergent African economic system as a whole.

This option deserves further investigation given the economic benefits that could emerge in die drive to meet SDG 1.

Author Biography

Anton Alberts is admitted as an advocate/barrister of the High Court of South Africa specialising in the legal fields of media law, ICT and space law. He is currently a Member of Parliament in South Africa and serves as a full member on the Parliamentary Portfolio Committee on Trade and Industry where he, amongst others, promotes the development of the country's space industry. He received his legal education at the University of Johannesburg where he obtained the degrees, B.A. (Law), LL.B, and LL.M (International Law (Cum Laude)), as well as an M.Phil in Futures Studies from the University of Stellenbosch. He is a prolific researcher and has published several legal works. Anton's focus is now increasingly on Space Law and its development for a new era of cooperation between government and private industry.

United Nations Fellowship Program "Drop Tower Experiment Series" (DropTES)—Hands-on Experience in Microgravity Research

Thorben Könemann

Abstract

In order to foster the technical and management skills of young students from different educational institutions, it is essential to provide them with as many hands-on experiences as possible. These tasks include, among others, creating technical drawings, programming, developing and performing experiments, and acquiring project management skills. Such practical training is not only important for the students' professional career, but also in their day-to-day life. This chapter introduces the United Nations Fellowship Program "Drop Tower Experiment Series" (DropTES). DropTES forms part of the Human Space Technology Initiative (HSTI) of the United Nations Office for Outer Space Affairs (UNOOSA). The Center of Applied Space Technology and Microgravity (ZARM) in Bremen, Germany hosts the fellowship program on behalf of UNOOSA with the support of the German Aerospace Center (DLR) Space Administration. Together all three institutions bring this annual fellowship program to life, opening it to student teams from research entities located in Member States of the United Nations. The DropTES Fellowship Program invites heads of research institutions or groups, university professors and postdoctoral researchers to promote the program amongst their respective students. The selected student team gets to conduct their own scientific or technological experiment under short-term conditions of weightlessness at the Bremen Drop Tower, ZARM's microgravity research facility. Each research team can consist of up to four Bachelor, Master and/or Ph.D. students, who must be endorsed by their academic supervisor (team leader–Prof./Ph.D.). Each year, the announcement of

Dr. Thorben Könemann—with contributions by UNOOSA, DLR Space Administration, and DropTES Research Teams.

T. Könemann (✉)
Center of Applied Space Technology and Microgravity (ZARM), University of Bremen, Bremen, Germany
e-mail: thorben.koenemann@zarm.uni-bremen.de

© Springer Nature Switzerland AG 2019   195
A. Froehlich (ed.), *Embedding Space in African Society*, Southern Space Studies,
https://doi.org/10.1007/978-3-030-06040-4_12

opportunity for DropTES is posted between September of the current year until the end of January of the following year on the United Nations Office for Outer Space Affairs (UNOOSA) website. Completed application forms are accepted by UNOOSA within this time-frame. The drop tower experiment series for the selected research team is scheduled each November and consists of four sponsored drops or catapult launches during one week at the Bremen Drop Tower. Each experiment series is accompanied by an on-site experiment integration, which takes place one week prior to the series week. During the entire experiment preparation and realization phase, the student team is technically assisted by experts from ZARM. A travel subsidy and on-site accommodation are guaranteed for the entire research team. The DropTES Fellowship Program is designed to contribute to the global promotion of space education and microgravity research, particularly to enhance relevant capacity-building activities in developing countries. All information required to participate in DropTES can be found on the fellowship program website of the United Nations Office for Outer Space Affairs (DropTES—Website—http://www.unoosa.org/oosa/en/ ourwork/psa/hsti/capacity-building/droptes.html). Currently, applications for the sixth cycle of DropTES (2018–2019) are open for submission. The following sections will provide an overview of all the DropTES cycles between 2013 and 2018 along with a description of the main characteristics of the Bremen Drop Tower.

1   DropTES—Projects (2013–2018)

The United Nations DropTES Fellowship Program started its first cycle of projects in 2013. Up to now, five cycles have successfully been completed at the Bremen Drop Tower. In short, it can be stated that all students who participated in previous drop tower experiment series were able to benefit in full from all aspects of this fellowship program.

By successfully launching their ambitious DropTES projects, the students were able to enhance both their knowledge and technical skills at the same time. Undoubtedly, they learned a lot from the program and made their first hands-on experience in microgravity research.

The following overview is a brief summary divided into subsections, containing further information about each single DropTES cycle. It is based on the abstracts of the submitted final reports from the selected research teams. This gives an overview on possible microgravity experiments within the scope of this fellowship program, and new ideas for future DropTES applications.

1.1 First DropTES Cycle (2013–2014)

Affiliation of Selected Research Team:

German Jordanian University, Jordan

Experiment Title:

"Tuned Mass Damping System for a Pendulum in Gravity and Microgravity Fields".[1]

Abstract written by Selected Research Team:

A tuned mass damper system, for short we refer to it as tilger, is suggested as damper of oscillations of tethers in satellites as they orbit the Earth while traversing the Earth's ionosphere and magnetic field. Variable Lorentz forces occur on the tethers, which will cause them to oscillate and may go out of control, which may de-orbit the satellite and fall to Earth. A system composed of a tuned mass damper and a simple pendulum simulating the tether is constructed. 350 sets of experimental trials were done on the system, while it is installed inside a drop tower capsule first resting on the ground, in order to pick four optimum setup experiments that will undergo a series of microgravity experiments at the Bremen Drop Tower in Bremen, Germany. We found that the oscillations of the simple pendulum will not be affected by the tilger during the free fall experiment, except if a feedback mechanism is installed between the simple pendulum and the tilger. In this case, the tilger will dampen the simple pendulum oscillations during free fall (Figs. 1 and 2).

1.2 Second DropTES Cycle (2014–2015)

Affiliation of Selected Research Team:

Universidad Católica Boliviana „San Pablo", Bolivia

Experiment Title:

"Performance and Mechanical Evaluation of the Bio Material Nitinol under Microgravity Conditions".[2,3]

---

[1]DropTES Final Report 2014—http://www.unoosa.org/documents/pdf/psa/hsti/DropTES/Final_Report.pdf.
[2]DropTES Final Report 2015—Part 1 (Nitinol)—http://www.unoosa.org/documents/pdf/psa/hsti/DropTES/Final_Report_1_second_cycle.pdf.
[3]DropTES Final Report 2015—Part 2 (Atrial Septal Defect/ Patent Ductus Arteriosus)—http://www.unoosa.org/documents/pdf/psa/hsti/DropTES/Final_Report_2_second_cycle.pdf.

Fig. 1 Experiment assembly and integration at the Bremen Drop Tower during the integration week. [© ZARM]

Abstract written by Selected Research Team:
[Part 1—Nitinol]

This article is intended to analyze changes in mechanical qualities of the alloy known as Nitinol (Nickel and Titanium) by means of two experiments realized under microgravity conditions compared to those of earth gravity. The first experiment is to analyze the angular velocity during the reconfiguration process of the material. On the other hand, the second experiment is to analyze the elasticity coefficient and rupture force capacity of the same material. The results were obtained at the Drop Tower, property of the ZARM (Center of Applied Space Technology and Microgravity) (Fig. 3).

[Part 2—Atrial Septal Defect (ASD)/Patent Ductus Arteriosus (PDA)]

This article is intended to analyze the performance regarding the occlusion of medical devices for congenital cardiopathies PDA-R and ASD-R through a cardiac emulator subjected to Microgravity conditions compared to the results obtained with earth gravity. The experiment is basically to comparatively determine the occlusion percentage of the aforementioned devices. Such results were obtained at the Drop Tower, property of ZARM (Center of Applied Space Technology and Microgravity).

Fig. 2 Integrated microgravity experiment of the first DropTES cycle inside the drop tower capsule—"Tuned Mass Damping System for a Pendulum in Gravity and Microgravity Fields". [© ZARM]

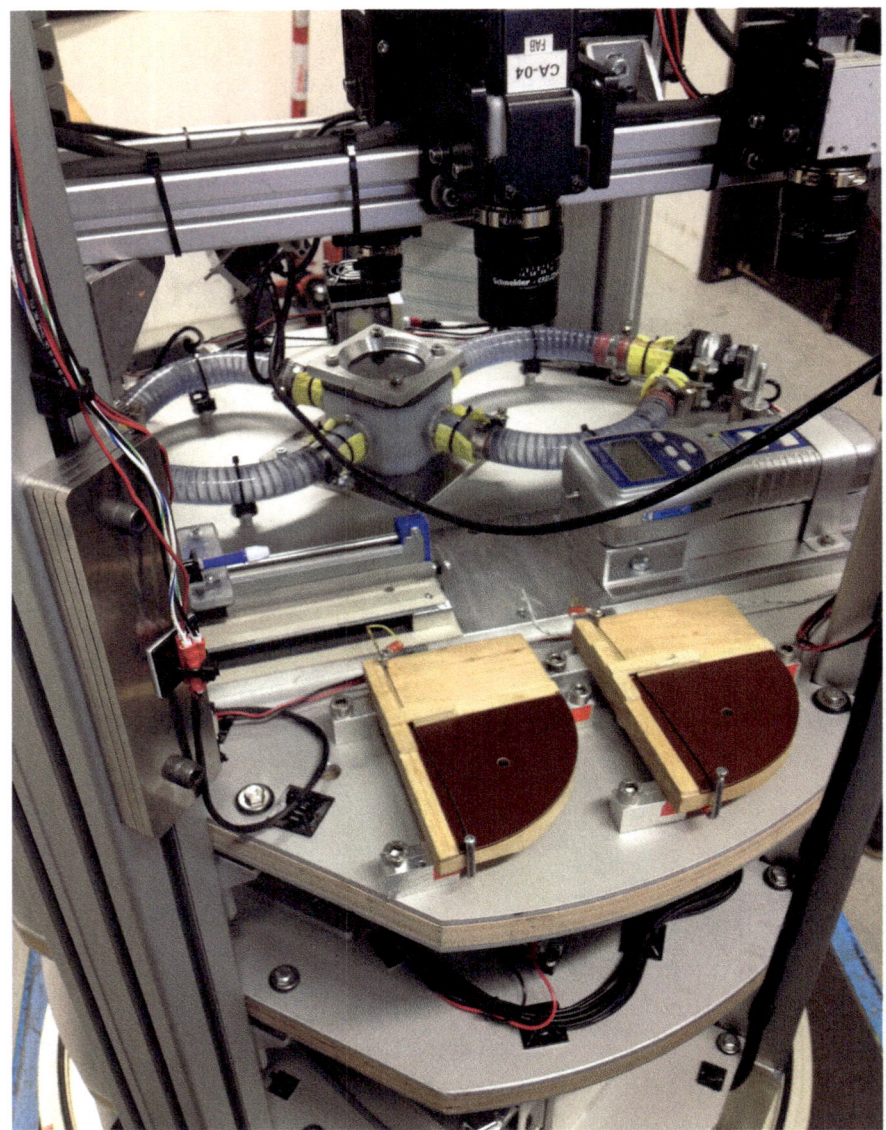

Fig. 3 Integrated microgravity experiment of the second DropTES cycle inside the drop tower capsule—"Performance and Mechanical Evaluation of the Bio Material Nitinol under Microgravity Conditions". [© ZARM]

1.3   Third DropTES Cycle (2015–2016)

Affiliations of Selected Research Team:

Instituto Tecnológico de Costa Rica (ITCR)/Universidad de Costa Rica (UCR), Costa Rica

Experiment Title:

"Behavior of a Reduced-Scale Robotic Arm Manipulator under Microgravity Conditions"[4]

Abstract written by Selected Research Team:

The general objective of the work was to design and implement an experiment in microgravity to analyze the influence of torsion, inertial, and reactive forces on a scaled robotic system, which is on a rotating, non-inertial frame of reference. For this experiment, the robotic manipulator was designed within the considerations and requirements of the Drop Tower at the Center of Applied Space Technology and Microgravity (ZARM). The implementation of the robotic manipulator prototype was carried out and then adapted to the drop tower chamber with centrifuge. With a predefined movement sequence, measurements of forces on the arm were carried out by means of load cell sensors and an inertial measurement unit (IMU), both in microgravity conditions and normal earth conditions. During the experiment series, four drops were successfully completed, gathering information from sensors and cameras recording the experiments. The results were analyzed and compared with respect to theoretical expectations and a conference paper publication was prepared, targeting a submission to the International Aeronautic Congress 2017 (Figs. 4 and 5).

1.4   Fourth DropTES Cycle (2016–2017)

Affiliation of Selected Research Team:

Warsaw University of Technology, Poland

Experiment Title:

"Experiment of Deployment of PW-Sat2 Satellite's Deorbit Sail in Micro-gravitational Conditions during Drop in the Bremen Drop Tower"[5,6]

---

[4]DropTES Final Report 2016—http://www.unoosa.org/documents/pdf/psa/hsti/DropTES/Final Report_3rd_cycle.pdf.
[5]DropTES 2017—Report on UNOOSA Website—http://www.unoosa.org/oosa/en/ourwork/psa/ hsti/capacity-building/droptes-fourth-cycle.html.
[6]PW-Sat2—Homepage—http://pw-sat.pl/en/home-page/.

Fig. 4 DropTES 2016—research team during experiment integration at the Bremen Drop Tower. [© ZARM]

Abstract written by Selected Research Team:

PW-Sat2 is a CubeSat project developed by more than 30 members of Students' Space Association (SKA) at the Faculty of Power and Aeronautical Engineering at Warsaw University of Technology (WUT), Poland. The project started in 2013 and it is scheduled for launch on Falcon 9 in late 2017. PW-Sat2 is a 2-unit CubeSat that carries several experiments on-board: 4m2 deorbit sail, custom Sun sensor, deployable solar panels and custom Electrical Power System (EPS).

The primary objective of PW-Sat2 is to test its main payload—the deorbit sail system and to verify its effectiveness. The PW-Sat2 deorbit system takes the form of a square sail (2 m sides). The material of the sail (6 µm thick aluminized Mylar film) is stretched across four flat springs, coiled around a specially shaped center core and placed in a cylinder with a diameter of 80 mm. The height of the entire system is less than 70 mm. After burning through a Dyneema wire, the sail will be unlocked and deployed by a stabilized conical spring to a safe distance (ca. 20 cm) away from the satellite. During the deployment procedure the flat springs of the sail structure will extend and assume their original C-shape which will stiffen the entire structure. As a result, the area and aerodynamic drag of the satellite will be significantly increased, accelerating deorbitation of the satellite (Fig. 6).

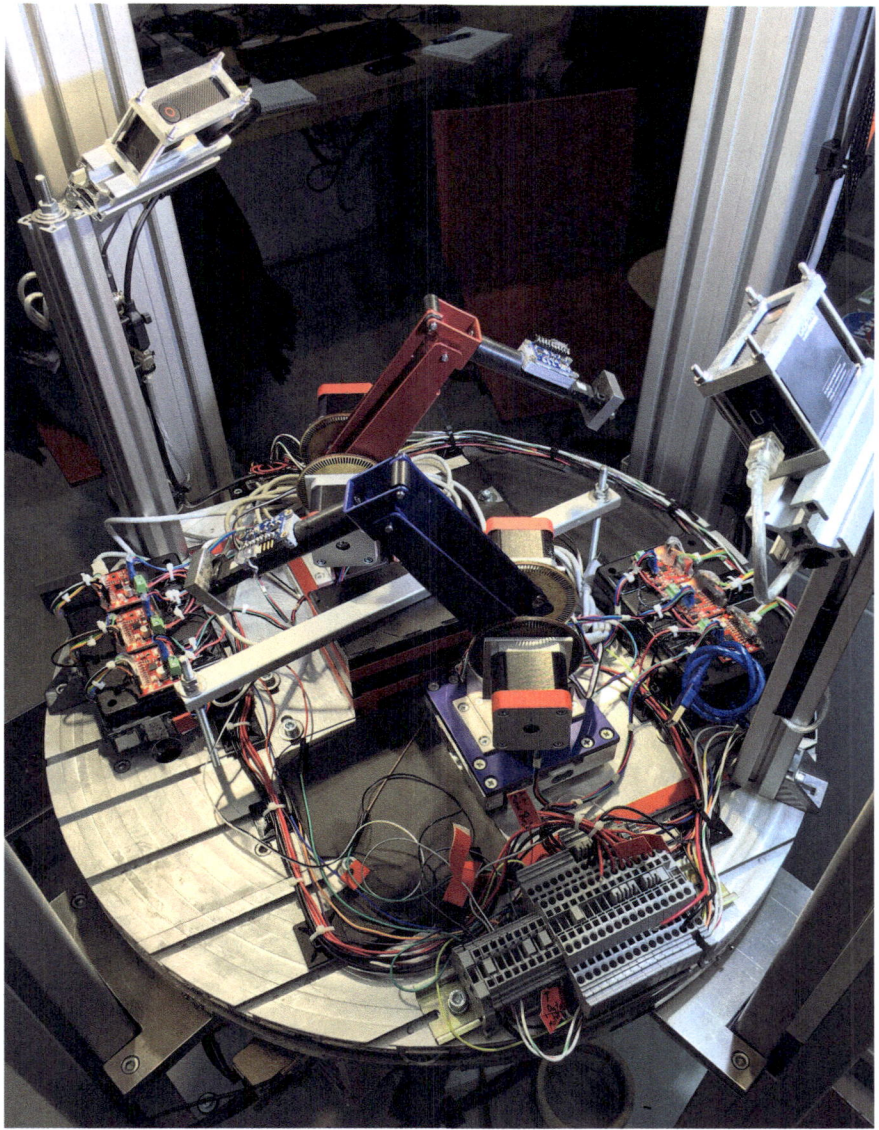

Fig. 5 Integrated microgravity experiment of the third DropTES cycle inside the drop tower capsule—"Behavior of a Reduced-Scale Robotic Arm Manipulator under Microgravity Conditions". [© ZARM]

(a)

Fig. 6 **a** (top)/**b** (bottom) Special integrations of equipment inside the deceleration chamber of the Bremen Drop Tower for the fourth DropTES cycle to observe sail deployments in vacuum conditions during free fall of PW-Sat2-CubeSat (test satellite)—"Experiment of Deployment of PW-Sat2 Satellite's Deorbit Sail in Micro-gravitational Conditions during Drop in the Bremen Drop Tower". [© ZARM]

(b)

Fig. 6 (continued)

1.5   Fifth DropTES Cycle (2017–2018)

Affiliations of Selected Research Team:

University of Bucharest (UB)/University Politehnica of Bucharest (UPB), Romania

Experiment Title:

"Interaction of Laser Exposed Medicine Droplets with Target Surfaces under Microgravity Conditions".[7]

Due to the fact that the fifth cycle of DropTES is ongoing whilst writing this chapter, further information will be available on the DropTES website of UNOOSA at a later stage.

2   ZARM—Bremen Drop Tower

The Center of Applied Space Technology and Microgravity (ZARM)[8] founded in 1985 is part of the Department of Production Engineering at the University of Bremen, Germany. ZARM is mainly concentrated on fundamental investigations of gravitational and space-related phenomena under conditions of weightlessness as well as questions and developments related to technologies for space. At ZARM, about 100 scientists, engineers, and administrative staff as well as many students from different disciplines are employed. Today, ZARM is one of the largest and most important research centers for space sciences and technologies in Europe.

The Bremen Drop Tower is the predominant facility of ZARM and also the only drop tower of its kind in Europe. ZARM's ground-based laboratory offers the opportunity for daily short-term experiments under conditions of high-quality weightlessness at a level of 10^{-6} g. Scientists from all over the world may have access to the drop tower facility and choose up to three times a day between a single drop experiment with 4.74 s in simple free fall and an experiment in ZARM's worldwide unique catapult system with 9.3 s in weightlessness.

Since the start of operation of the drop tower facility in 1990, thousands of drops/catapult launches based on hundreds of different experiment types from various scientific fields like fundamental physics, combustion, fluid dynamics, planetary formation/astrophysics, biology, and material sciences have been accomplished so far. In addition, more and more technology tests have been conducted under microgravity conditions at the Bremen Drop Tower, in order to prepare appropriate space missions in advance.

---

[7]DropTES 2018—Proposal.
[8]http://www.zarm.uni-bremen.de.

Fig. 7 The Bremen Drop Tower with a total height of 146 m located on the campus of the University of Bremen (technology park) in Germany. [© ZARM]

As part of its educational purposes ZARM contributes to a variety of student programs like "Drop Your Thesis!"[9] by the European Space Agency (ESA)—Education Office, "DropTES" (Drop Tower Experiment Series)[10] by the United Nations Office for Outer Space Affairs (UNOOSA) and the German Aerospace Center (DLR), and the German-Swedish Student Program "REXUS/BEXUS" (Rocket/Balloon Experiments for University Students)[11] by DLR and the Swedish National Space Board (SNSB).

---

[9]http://www.esa.int/Education/Drop_Your_Thesis/About_Drop_Your_Thesis.
[10]http://www.unoosa.org/oosa/en/ourwork/psa/hsti/capacity-building/droptes.html.
[11]http://rexusbexus.net.

2.1 Drop Tower Operation

In February 1987, the planning and construction period of the Bremen Drop Tower began and it was completed within only three years. With a total height of 146 m and its characteristic glass roof the drop tower was manufactured of a steal vacuum tube with a height of 122 m which is enclosed by a concrete tower (Fig. 7). In September 1990, the ZARM Drop Tower Operation and Service Company (ZARM FAB mbH) owned by the State Government of Bremen was established as a public company operating and maintaining the drop tower facility simultaneously with the time of beginning microgravity experiment operations. Up to now, ZARM's drop tower is the only ground-based facility of this kind in Europe for short-term experiments under conditions of high-quality weightlessness. Furthermore, ZARM's drop tower has become one of the most important and outstanding microgravity facilities worldwide.

The high-quality weightlessness achieved for microgravity experiments at the drop tower in Bremen is of the order of 10^{-6} g which is nearly comparable to drag-free satellites. Such an excellent microgravity quality exceeds that of (un-) manned sub- or orbital platforms (e.g. even ISS) by orders of magnitude. Moreover, the Bremen Drop Tower represents a very economical alternative with a permanent access to weightlessness on earth.

At the drop tower facility, the free-fall duration of a microgravity experiment is about 4.74 s given by a height of the inner cylindrical drop tower vacuum tube of 122 m (110 m drop distance). In December 2004, the world-wide unique capsule catapult system developed by ZARM has started its operation of microgravity catapult experiments. During a catapult operation the experiment capsule performs a vertical parabolic flight inside the vacuum tube of the drop tower. This way, the condition of weightlessness can be extended to 9.3 s at which the same high quality of microgravity is achieved like for the simple drop mode.

The repetition rates for both the drop and the catapult operations are the same. Up to 3 drops or catapult launches can be performed per day. A combined system of 18 vacuum pumps requires about 1.5 h for the evacuation of the drop tower vacuum tube with a volume of more than 1700 m^3. Such a complex multistage pump system is necessary to achieve a residual pressure of 10 Pa realizing the excellent conditions of weightlessness—minimal air drag on the freely falling experiment capsule. For the recovery of the capsule, the vacuum chamber has to be re-flooded with air within 30 min. Afterwards, the experiment is again available for the scientists for further investigations at the working area in the integration hall (Fig. 8) or in a provided laboratory of ZARM.

2.2 Catapult System

Located below the base of the drop tower, a chamber of 11 m depth contains the catapult system (Fig. 9). During the catapult operation the experiment capsule performs a vertical parabolic flight inside the drop tower vacuum tube. This meets

Fig. 8 Experiment capsule integration areas at the main machine hall of ZARM's drop tower facility. [© ZARM]

scientist's demands on extending the experiment time to 9.3 s at the same high quality of microgravity (10^{-6} g).

The catapult capsule (Fig. 10) is accelerated by a pneumatic piston driven due to the pressure difference between the vacuum of the evacuated drop tower tube and the pressure of the pressurized air tanks of the catapult system. The capsule acceleration level is adjusted by means of a servo hydraulic braking system controlling the catapult piston velocity. The deceleration container inside the vacuum chamber is moveable to fall back in the capsule capture position within 3 s after the lift-off and passing of the catapult capsule.

ZARM's worldwide unique catapult system is currently able to accelerate catapult capsules with gross weights up to the actual limit of 400 kg (an enhancement of up to 500 kg is available on special request) to a lift-off speed of 46.9 m/s within 0.28 s.

2.3 Experiment Operation

The microgravity laboratory system consists of a cylindrical capsule with an outer diameter of 814 mm and a total length of 2094 mm for the short drop/catapult capsule and 2860 mm for the long drop capsule (Fig. 11). Inserted platforms bordered by aluminum profiles form the modular capsule structure. The number of platforms depends on the space required for experimental studies. The standard equipment of each experiment capsule includes a computer platform as well as the accumulator platform necessary for the internal power supply (nominal 24 V/at charging 28 V DC, max. 40 A, 25 Ah). The Capsule Control System (CCS) of the drop or catapult capsule performs the experimental control and data storage, as well as interactive experimental regulation through telemetry during the test on ground and the following experiment phase in microgravity. After the integration of the experiment, the capsule will be closed pressure-tight with an aluminum cover.

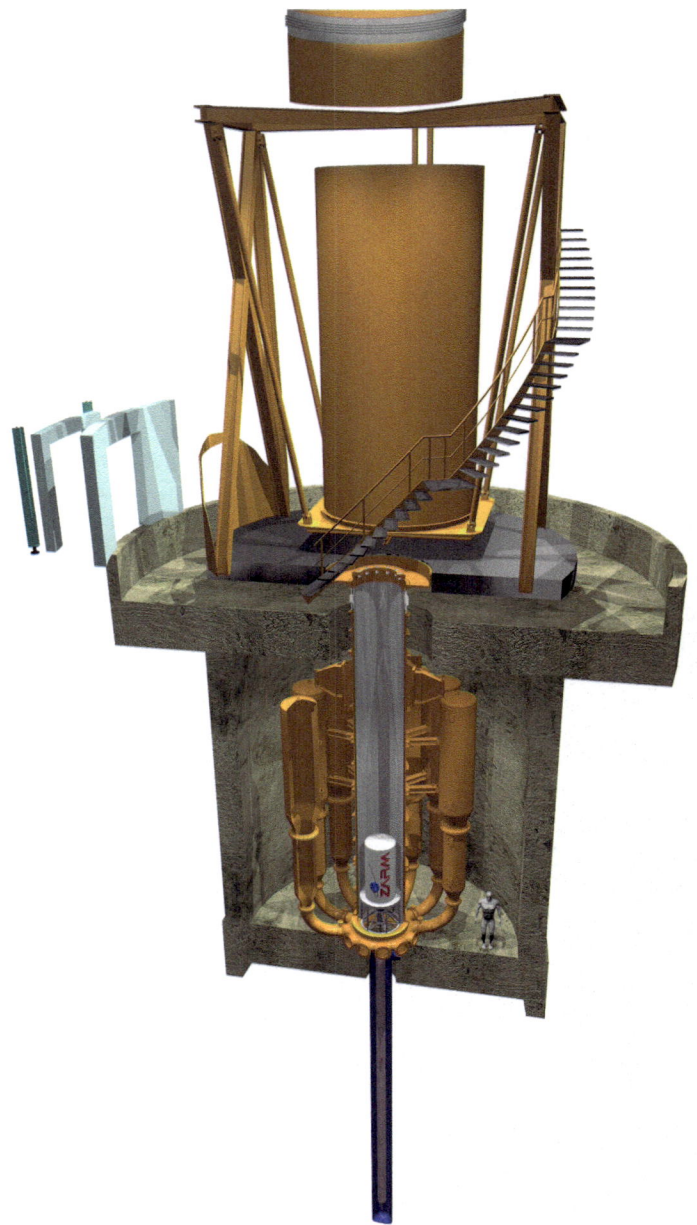

Fig. 9 Scheme of the catapult system of the Bremen Drop Tower including the catapult capsule on the pneumatic piston, the pressurized air tanks, and the moveable deceleration container on the top. [© ZARM]

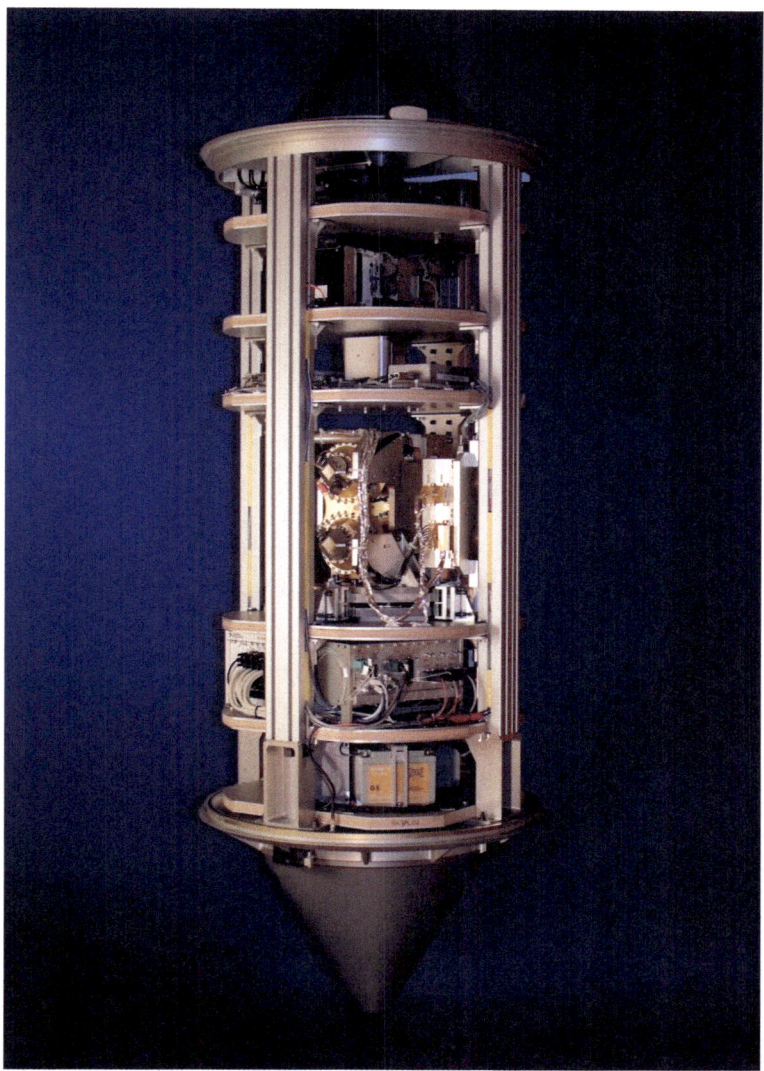

Fig. 10 Example of a fully assembled catapult capsule incl. microgravity experiment, Capsule Control System, and battery pack. [© ZARM]

The inner steel vacuum tube of the drop tower has no connection to the outer concrete tower. The pedestal of the vacuum tube is fixed to the lower positioned deceleration chamber with a height of 13 m and it is eccentrically arranged to the vertical axis of the concrete shell. By this, one can avoid the transfer of wind-induced tower oscillations to the sensitive experimental operation to the greatest possible extent. Depending on the experiment campaign, for a drop with

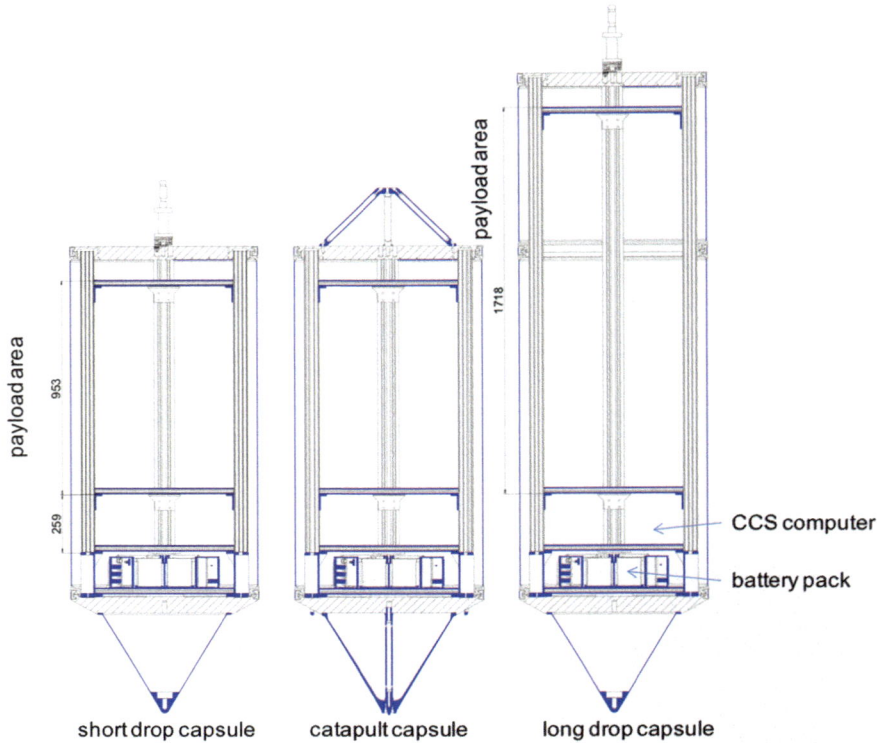

short drop capsule   catapult capsule   long drop capsule

Version	short	long	catapult
stringer lenght [mm]	1341	2107	1341
max. payload height [mm]	953	1718	953
total area of experiment-platform [m²]		0.359	
base structure incl. batteries and CCS [kg]		122.8	
top lid plate incl. video transmission unit [kg]		32.9	30.2
pressurized cover incl. clamping rings and thermal isolation cover [kg]	38.5	61.1	38.5
4 stringer [kg]	36.8	57.4	36.8
nose cone incl. connection rod [kg]		4.6	10.2
1 experiment-platform incl. brackets [kg]		15.5	
capsule net weight [kg]	235.6	278.8	238.5
capsule gross weight [kg]		500	400*
max. payload mass [kg]	264.4	221.2	161.5

* actual limit, enhancement up to 500 kg in
future depends on evolution progress

Fig. 11 Experiment capsule types of the Bremen Drop Tower. [© ZARM]

4.74 s of microgravity the capsule must be brought up to a height of 120 m by a winch, for the catapult operation with 9.3 s of microgravity the capsule must be placed on the catapult piston pulled down to 11 m below the base of the drop tower (Fig. 12).

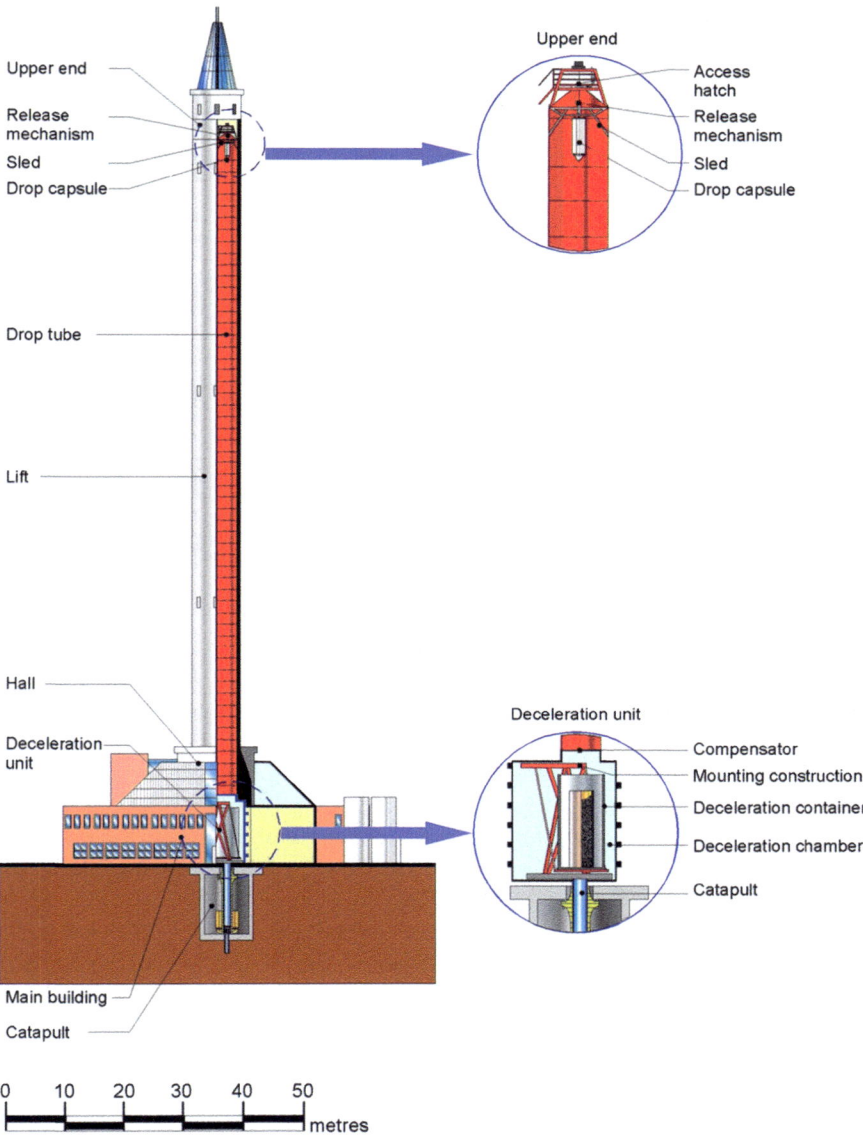

Fig. 12 Scheme of ZARM' drop tower facility including drop release, catapult, and deceleration mechanism. [© ZARM]

Fig. 13 Ground control station of the drop tower. [© ZARM]

Then, the inner steel tube of the drop tower must be closed pressure-tight so that the evacuation of the 1700 m^3 capacity of the drop tower and the deceleration chamber can start—this is necessary to minimize the effect of the air drag on the freely falling capsule. 18 combined vacuum pumps require about 1.5 h for the evacuation of the steel tube. The capsule is then dropped or launched at a residual pressure of 10 Pa. For this, the final launch command will be given by the scientist from the ground control station of the drop tower (Fig. 13). As soon as the initial disturbances caused by the drop release or catapult launch have been damped down, residual accelerations of 10^{-6} g can be detected during the free fall from 110 m for a drop experiment or during the vertical parabolic flight of a catapult experiment.

The drop capsule arriving at the braking zone with a final speed of 167 km/h and the catapult capsule achieving a lift-off speed of 175 km/h are gently stopped in the deceleration container. The container is filled with fine polystyrene pellets. The nose cone of the capsule reduces the entry peak and stabilizes the vertical axis during braking. Here, a maximum deceleration of 50 g acts on the incoming capsule (Fig. 14). The duration of the impact of the capsule is about 200 ms. During a catapult launch with a total duration of about 300 ms, the catapult experiment is accelerated up to 30 g (Fig. 15). The acting acceleration force on the catapult capsule is very smooth due to a sinus curve velocity profile of the catapult piston movement.

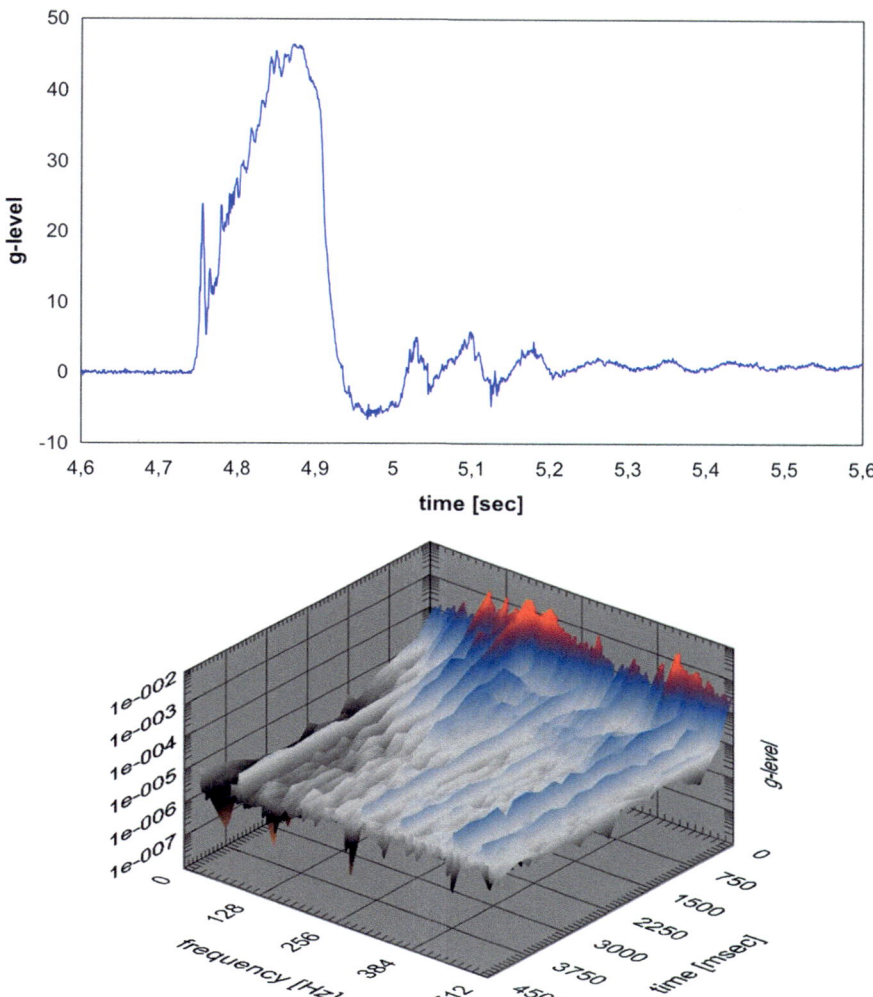

Fig. 14 Typical capsule deceleration plot. Here, during drop operation (top). Waterfall amplitude spectrum of drop axis (bottom). [© ZARM]

To recover the capsule from the deceleration container, the vacuum chamber (Fig. 16) must be first re-flooded with preconditioned air—this takes about 30 min. Afterwards, the scientists regain the capsule with the microgravity experiment at the integration area inside the machine hall or in an air-conditioned laboratory if requested.

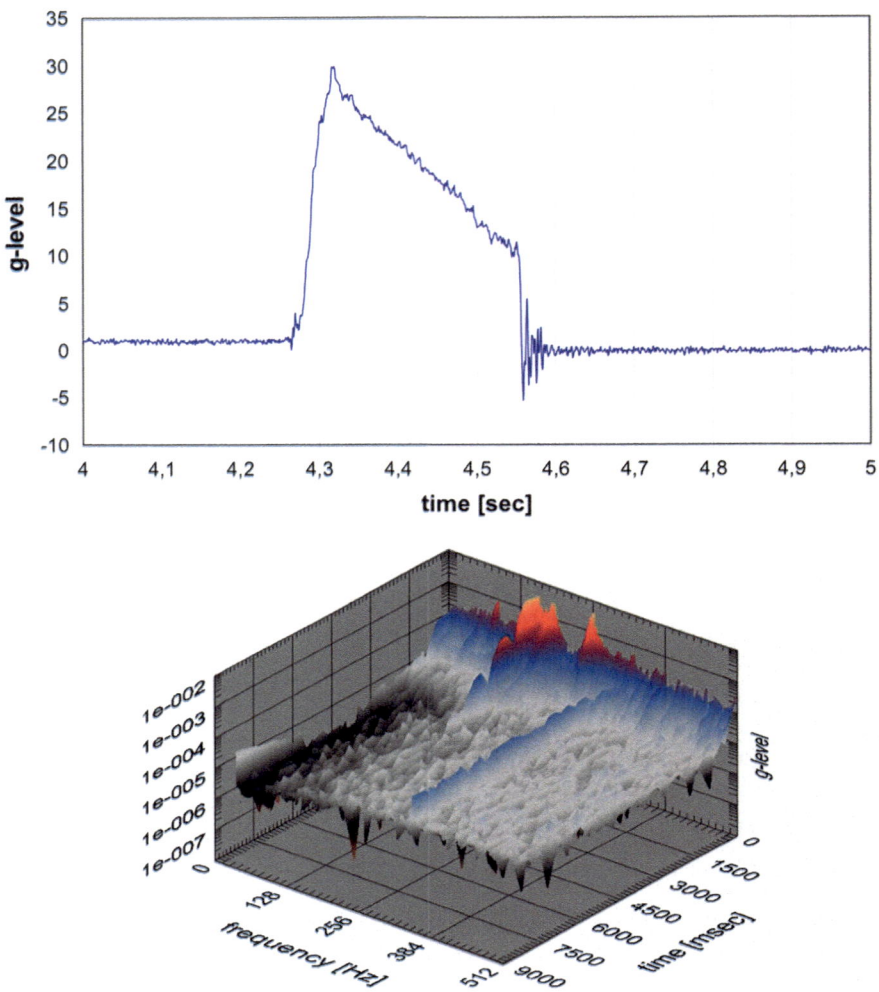

Fig. 15 Typical catapult capsule acceleration plot. Neglect the delay time of about 4 s after initialization of launch (top). Waterfall amplitude spectrum of catapult axis (bottom). [© ZARM]

2.4 Standard Capsule Equipment

The Capsule Control System (CCS) located at the bottom of the capsules is part of our standard capsule equipment that operates the drop or catapult experiments via remote control. The CCS is based on a National Instruments™ PXI Chassis 1000B-DC with a Real Time Controller 8145 RT containing a PXI-6031E (Dev1), a PXI-6527E (Dev2), a PXI-6713 (Dev3), and a second PXI-6031E (Dev4) device. The connection of experiment's analogue and digital I/O channels to the CCS is performed via a separate interface board (Fig. 17).

Fig. 16 Capsule recovery after experiment operation—(left) capsule deceleration container at the bottom of the vacuum tube of the Bremen Drop Tower. [© ZARM]

In order to control the experiment hardware during the drop tower operation, each part of the I/O channels can be displayed and managed on a LabView™ screen in the control station on the ground floor. All data acquired during the experiment will be continuously transmitted via WiFi to the ground control and saved to the password-secure user account.

As a further standard equipment of our drop tower capsules, the Power Distribution Unit (PDU) provides six switchable current channels supplied with power by a rechargeable battery pack for the experiment operation during free fall (nominal 24 V/max. 40 A, 25 Ah—continuously buffered by an external power supply (28 V/10 A) in laboratory and until approx. 1.5 min prior to the launch command). The standard on-board sensor pack (further capsule sensors are available on request) includes:

- *acceleration—three axes (range:± 1 g)*
- *deceleration—capsule axis (range:± 50 g)*
- *capsule pressure (nominal pressure: 1.013 hPa)*
- *capsule temperature (in general: room temperature)*

All sensor values are monitored during drop tower operation and stored on the password-secure user account.

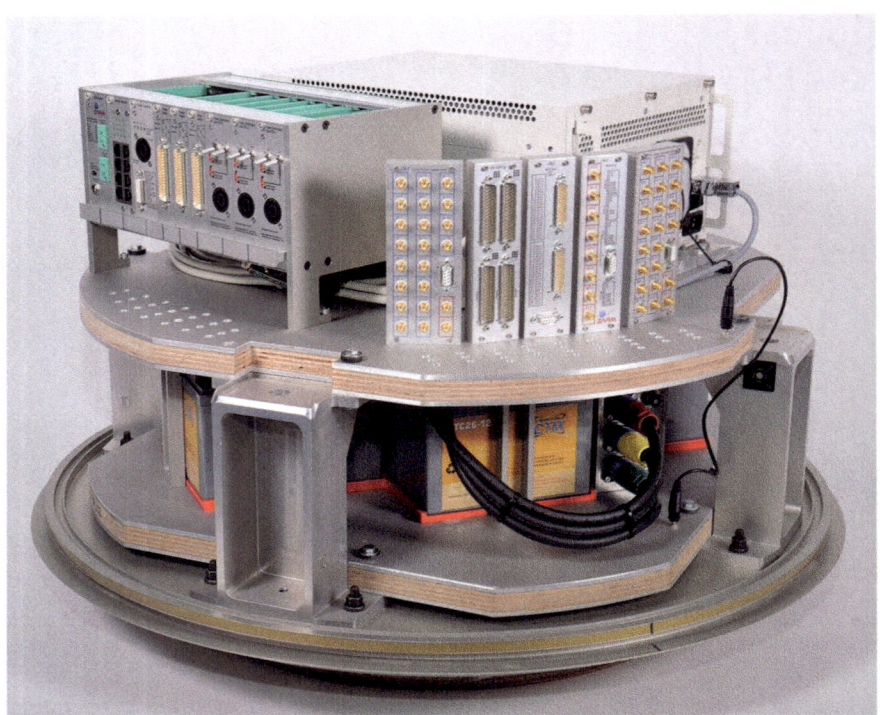

Fig. 17 National Instruments™ Real Time Capsule Control System (CCS)—(top). Experiment interface board of the CCS (SMA- and D-Sub-connectors as a standard plug)—(bottom). [© ZARM]

Fig. 18 On-board microgravity centrifuge incl. partial gravity experiment installed into the long drop capsule structure

2.5   Special Capsule Equipment

On request, the ZARM Drop Tower Operation and Service Company (ZARM FAB mbH) provides a digital high-performance video system (Photron Fastcam MC2™—2 GB) and problem-specific lenses to all drop or catapult experiments. This high-speed video system is based on light sensitive monochrome or optionally color CMOS imaging sensors (Model 10 K; 512×512 pixel resolution) and its live video output can be transmitted to the ground control prior to and during the experiment. The transmitted data has a bandwidth of standard video and the high-speed data is directly stored onboard at a chosen frame rate of up to 10,000 fps.

If required, all experiments can be connected to a thermal liquid heating and cooling circuit. This liquid circuit (glycol/water-mixture) is linked to a thermostat outside of the drop tower vacuum tube. Through a closed loop regulation, the temperature can be adjusted between -20 °C and $+60$ °C. The liquid circuit will be disconnected approx. 1.5 min prior to the launch command. In addition, we offer the option of a non-standard power supply for both drop and catapult experiments. This external power supply provides an adjustable DC voltage with up to 100 A and will be also disconnected from the capsule lid plate approx. 1.5 min prior to the launch command.

The release of gases during the experiment (e.g. from cryogenic devices or combustion exhausts) is regulated/served by a vent line. Its connectors are located on top of the capsule's lid plate (alternative use of the connectors of the heating and cooling circuit up to approx. 1.5 min prior to the launch command). The gases can be either released outside of the drop tower tube or directly to the ambient vacuum. To avoid thruster effects during the free fall, the vent line must be closed prior to the drop or launch of the capsule and the gases must be stored in on-board containers.

For experiments which request data from accelerations between 1 and 0 g we recommend the application of a specially designed on-board microgravity centrifuge (Fig. 18). Basically, this centrifuge consists of a rotating platform equipped with a number of slip-ring transducers for the supply with electrical power and signal transmission between rotating platform and capsule. The on-board microgravity centrifuge is not applicable for catapult operation due to safety aspects.

Author Biography

Dr. Thorben Könemann is the Deputy Scientific Director of the ZARM Drop Tower Operation and Service Company (ZARM FAB mbH) located in Bremen, Germany. He is mainly responsible for consulting the scientific customers and managing the overall microgravity experiment operation. Furthermore, he is currently the vice-chair of the Scientific Commission G—Material Sciences in Space—of the Committee on Space Research (COSPAR). During his student time at ZARM, he was involved in the MICROSCOPE project for a short period. In 2006, he finished his diploma thesis in the field of quantum physics and immediately proceeded his work as a Ph.D. student in the same project at ZARM. Within this pilot project called QUANTUS (Quantum

Systems under Weightlessness), Dr. Könemann was a member of the team that realized the world-wide first Bose-Einstein Condensate under conditions of weightlessness at the Bremen Drop Tower. Up to now, the QUANTUS project and its follow-on projects represent an emerging area of science in quantum engineering with an impressive potential for future technology developments and multidisciplinary applications. In many proceedings and publications including one remarkable publication in Science Magazine—Science 328, 1540 (2010)—related to this topic, Dr. Könemann made appropriate contributions as author as well as co-author.

The European Space Agency (ESA) and the United Nations 2030 SDG Goals

Isabelle Duvaux-Béchon

Abstract

Since years, space activities, *inter alia* the ones of the European Space Agency (ESA) are an important tool serving development. In the frame of the corporate "Space for Earth" initiative aiming at presenting the ESA projects, services, applications or technologies along thematic or regional approaches facing similar challenges, specific emphasis has been put since 2016 on sustainable development as one of the main challenges faced on Earth. The UN 17 Sustainable Development Goals (SDG) adopted in September 2015 are benefiting and can benefit much more from space tools, either for the monitoring of goals or for the support to the achievement of goals. In order to ease the identification of the relevant activities by potential users, ESA developed a catalogue listing its projects that can support one or more of the goals. The first on-line version is available since 15 March 2018. It will be regularly expanded with new ESA activities and activities from ESA partners. Discussions with the UN Office of Outer Space Affairs (OOSA) in the frame of the preparation of UNISPACE+50 led to the signature of a Joint Statement for further cooperation declaring inter alia the intent of the two Organisations to do a joint development of the ESA Catalogue and the OOSA Compendium. Further to the request of several ESA Member States, Africa has been agreed as a regional area for implementation of the support to the SDGs. In particular, the catalogue will be used to identify the existing space activities relevant for Africa (already done in Africa or that can be implemented also in Africa). The next steps of this initiative will include the expansion of the catalogue and the identification of "users" that could benefit from "space", the inclusion of information concerning

I. Duvaux-Béchon (✉)
Head of the Member States Relations & Partnerships Office,
European Space Agency (ESA), Paris, France
e-mail: Isabelle.duvaux-bechon@esa.int

© Springer Nature Switzerland AG 2019
A. Froehlich (ed.), *Embedding Space in African Society*, Southern Space Studies,
https://doi.org/10.1007/978-3-030-06040-4_13

socio-economic benefits of the space activities, and the extension of the catalogue to the activities of other entities than ESA.

1 Introduction

At the time the UN Sustainable Development Goals (SDG) were adopted, ESA was already in a process of coordinating its activities in support of sustainable development, as it was recognised since years that space activities are indeed an important tool serving development. It was thus natural to link the two initiatives, as a way to be closer to the concerns of the "users" by making reference to goals that are recognised by all and neutral vis-à-vis the space sector, the SDGs. The concept behind such initiative by ESA and the resulting catalogue that has been elaborated, are presented in this paper, together with some next steps.

2 The European Space Agency (ESA)

2.1 ESA in Short

The European Space Agency (ESA) is an inter-governmental organisation with today 22 Member States and several Associate and Cooperating States. According to the Article 2 of its Convention, it has been established "to provide for and promote, for exclusively peaceful purposes, cooperation among European states in space research and technology and their space applications". With a budget above 5 B€ per year, coming mainly from its Member States and the European Commission, ESA is in charge of the development of space projects, across all domains, including Earth Observation, Telecommunications, Navigation, Human Spaceflight, Science, Technology or Launchers.

The 22 Member States have each one vote in the ESA Council. They agree on the projects to be undertaken and decide on their contributions (compulsory and according to the GNP for Science and core transverse activities, on a purely voluntary basis for all the other activities). ESA develops a certain number of applications, usually up to the proof of concept, and they are then transferred when operational to organisations or private companies.

ESA also put in place Technology Transfer and Business Incubation Centres to foster the intake of space technology and applications, and use of the space data, in society.

2.2 The Three Core Goals

ESA has three core goals for the space activities aimed at:

- Maximising the integration of space into European society and economy;
- Fostering a globally competitive European space sector;
- Ensuring European autonomy in accessing and using space in a safe and secure environment.

In the rest of the paper, we will concentrate on some aspects of the first goal, the background or reasoning behind being that projects developed by public money have to benefit as much as possible to the citizens and inhabitants of the world, via in particular the policies of the ESA and EU Member States that can be supported. "Space", whether it is for the technologies, the data, the applications or the services, can be considered as a tool supporting most challenges that are faced by people and societies on Earth, tools that should be better known and further developed.

And, in order to make it better known, a specific initiative was started some years ago and called "space for earth". Its aim is to present ESA activities not along the traditional programmes (Earth Observation, Telecommunications...) but along thematic or regional approaches in a matrix-like approach. Web pages have been developed along some of the thematic and regional challenges and are fed with news items from all the ESA projects that are relevant to those challenges.[1] The concept is not new as many of the ESA projects are already supporting this approach, however it is here tackled at the corporate level in order to help the external potential users to have one entry point to all ESA activities, across all Directorates.

3 Serving Major Earth Challenges

3.1 Transversal Approaches

With the identification of those challenges faced on Earth that can be supported by space technologies, data, applications or services, better service can be provided to those on Earth who could benefit from those projects. Indeed, many potential users do not even know that space is serving and how space can serve better. They might be aware of the largest projects without having the faintest idea of what it brings on Earth.

What is important is to make the links between those challenges and the concerned space projects to help the information pass both ways.

Two large categories of challenges were identified: the ones faced within a given geographical area, and the ones faced by a thematic topic. Even if the challenges are different, the approaches are quite similar, in a transversal way, in order to maximise this "integration of space into European society and economy".

In both approaches, a certain number of steps are structuring the initiatives:

---

[1]www.esa.int/spaceforearth and @spaceforearth.

- Identify the needs as expressed by Member States or main stakeholders
- Identify the main actors (at government or regional level, local level, NGOs, industry, international organisations…)
- Identify what is already done at ESA in these areas
- Ensuring the information is flowing both ways on what is being/can be done within the existing programmes
- Prepare the future with the identification of gaps in the fulfilment of the needs and for the definition of future programmes/activities to be proposed to Member States
- Build partnerships with non-space entities for a more efficient discussion and preparation of the future.

These two categories of challenges are recognised by ESA Member States, as requesting specific coordination efforts for better efficiency as can be read in the "Resolution on ESA programmes: outlook and way forward" adopted at the 2016 ESA Council at Ministerial level where:

> […] the Council, meeting at ministerial level, […] STRESSES the need for the Agency to take steps to contribute decisively, through a coordinated effort making use of data generated across ESA programmes, in meeting the major challenges faced worldwide by our societies, particularly sustainable development, bearing in mind the 2015 United Nations Sustainable Development Goals, better knowledge of the environment and economic development of specific areas while ensuring the protection of the corresponding valued ecosystems, such as the Arctic, the Atlantic Ocean and the Alpine region. […]

It is worth to be noted that all space domains contribute one way or another to one or more of the SDGs, from Earth observation, telecommunications, positioning, technology or Human spaceflight to operations. Space cannot solve all challenges on Earth and be "the" answer to development, but space projects, technologies, applications and services can be very valuable contributors, much more than now. Both for the monitoring of the indicators, and for support to the achievement of the goals.

Both approaches, "specific areas" and "sustainable development", through the UN Sustainable Development Goals (SDG) will be addressed in the next paragraphs.

3.2 Specific Areas: The Regional Approach

Since now a certain number of years, ESA, via some of its projects, is developing activities in support of challenges faced by Member States across consistent regional areas. It can concern regional areas embedded into the territories of ESA Member States and of specific interest, or areas not embedded, the economic development of which they are supporting. Each of the areas identified here has specificity that can make it a model for other similar areas that could benefit from the same space support.

When several Programme Directorates at ESA are supporting or could support the same regional area, of interest to several Member States, it is more efficient to develop a coordinated approach across ESA to maximise the identification of the needs and the support provided, and to minimise the resources needed to interface with the external world.

Those geographical areas have been chosen because:

- They are facing specific and difficult challenges (in some cases similar or shared) and several Member States are supporting a coordinated approach
- They can benefit from space assets from different domain areas and require tailored approaches and dedicated efforts.

In line with this approach, a certain number of geographical areas, supported individually by several Directorates if not all, are coordinated at corporate level.

ESA projects involved can include Earth Observation, Telecommunications, Navigation, Human Spaceflight and associated applications and technologies. They will support activities and policies linked to economic development, safety and security of people, infrastructures or resources, better living conditions for the population, border and maritime surveillance, better knowledge of environment or climate change, and much more.

In some cases, the establishment of partnership agreements with stakeholders has allowed or will allow understanding and consolidating the user requirements relevant to them and issuing specific calls/tenders for gap filling.

The regional areas under corporate coordination today include the 5A (Arctic, Antarctica, Atlantic, Alps and Africa) as well as Maritime (with focus on the Mediterranean). Several of them have challenges in common and this is also coordinated (Fig. 1).

3.3 Sustainable Development: ESA and the UN SDGs

3.3.1 Why Supporting the SDGs

Since 2008, ESA started to address a thematic challenge in a coordinated way, "energy", and a dedicated web page was built under the "Space for Earth" umbrella.[2]

When the United Nations adopted the 17 SDGs on 25 September 2015,[3] ESA was already in a process to develop a specific action towards "sustainable development"[4] with the recognition that a lot can be achieved via the help of "space". It was thus natural to build on these new Goals as they were reflecting the challenges (via the 169 targets that were defined) that the Nations of the world wanted to solve with the legitimacy that they have. It is thus not ESA that sets targets of the

---

[2]http://www.esa.int/Our_Activities/Preparing_for_the_Future/Space_for_Earth/Energy.
[3]https://www.un.org/sustainabledevelopment/sustainable-development-goals/.
[4]https://www.esa.int/SDG.

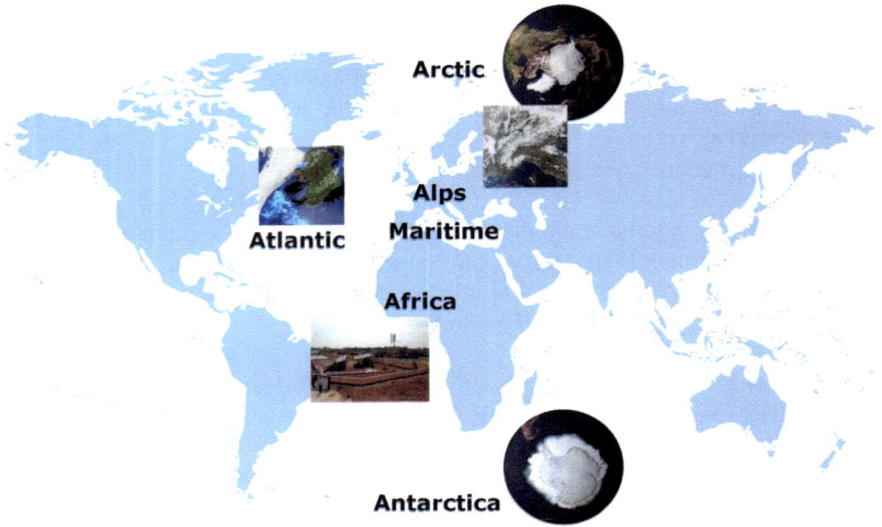

Fig. 1 The "5A" regional approach

challenges that should be addressed, but the Nations and ESA and other space agencies in the world that works towards supporting this process. Those Goals embed also all thematic challenges that can be addressed (the "Energy" challenge falls under SDG7, the "Health" challenge addressed more recently under the SDG3, and so on). An important feature of the SDGs,[5] and contrary to the former Millenium Development Goals, is that they are applicable to all countries, developed or developing, and have a universal value (Fig. 2).

When analysing those 17 SDGs, it appeared quite clearly that "space" is already helping, and could help much more, and is becoming an indispensable tool for many of the challenges. It can support the monitoring of the Goals and their targets (many variables can only be measured from space), taking the "temperature" of the planet and human activities, and it can support the achievement of the Goals, by helping improving the results achieved, as would "pharmaceuticals" do (Fig. 3).

3.3.2   The Catalogue of ESA Activities Supporting the UN SDGs

With this recognition and understanding internally and externally that all SDGs can get some support from space projects, technologies, applications or services, ESA Member States asked in June 2016 the Executive to build, as a proof of concept and a way to better inform the external world, a catalogue of ESA activities supporting the UN SDGs, and to enter into discussion with the UN to see how it would be possible to support the Goals in a more systematic and organised way.

---

[5]https://sustainabledevelopment.un.org/sdgs.

Fig. 2 The UN 17 Sustainable Development Goals

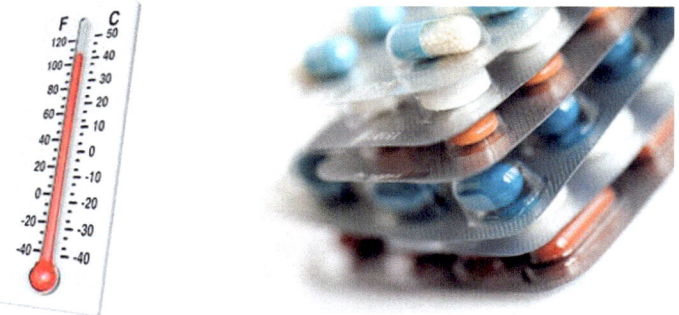

Fig. 3 Monitoring and achieving the Sustainable Development Goals

This was further confirmed by the first resolution adopted during the ESA Council at Ministerial level end 2016, resolution entitled "Towards Space 4.0 for a United Space in Europe" where Member States recognized that space serves:

[…] "societal needs, responds to European and global challenges and offers opportunities, notably those related to the attainment of Sustainable Development Goals, … mitigation of risks, … climate change…", […].

Work started immediately after June with a first demonstration catalogue end November 2016.[6] The online version with an initial set of some 300 activities went

---

[6]http://esamultimedia.esa.int/docs/spaceforearth/SD_Catalogue_COMPLETE_161128.pdf.

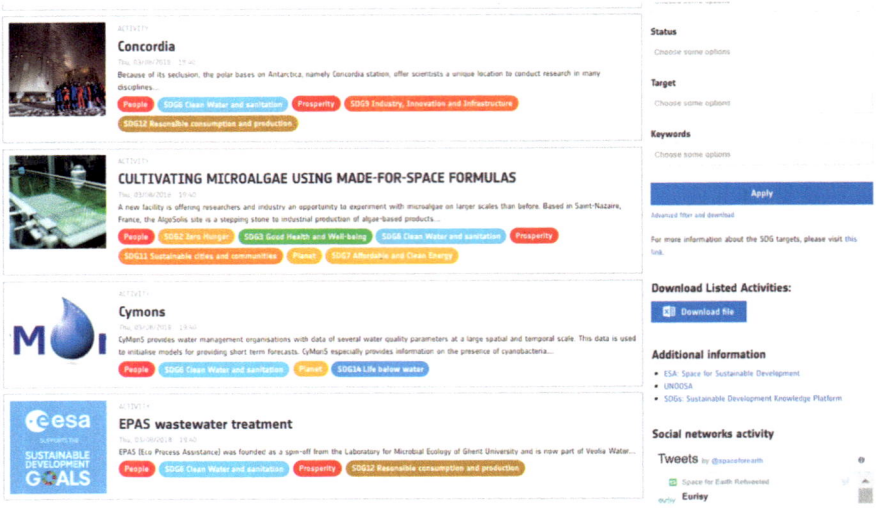

Fig. 4 The catalogue of ESA activities supporting the UN SDGs

live on 15 March 2018.[7] It will be regularly complemented as not all existing activities are yet in and all future activities will have to be included when mature (Fig. 4). The aim of the catalogue is to support potential users to identify the projects and activities that could help them achieving the challenges they are facing or the policies they have to implement. Projects can be searchable by SDG and target (e.g. SDG3 "Good health and well-being" and target 3.9 "By 2030, substantially reduce the number of deaths and illnesses from hazardous chemicals and air, water and soil pollution and contamination"), going out of the traditional "programme" approach. For each example, a short description is provided together with the status of availability as well as the link to the project page and contact points to go further.

Several hundreds of activities have been identified across all ESA Directorates. These are all existing or under development activities already financed by Member States.

Activities from all domains are included as they can all contribute, one way or another, to support sustainable development on Earth. For example, Earth observation data can help monitor the targets (climate evolution, temperatures...), and also support achievement of some goals, for example supporting the identification of good locations for the implementation of renewable energy. Telecommunications satellites in addition to lowering the digital divide, allow the development and provision of applications and services in remote areas. European navigation satellites provide independent precise positioning for efficient navigation of all mobiles and can be combined with other data for improved services. Human spaceflight and

---

[7]https://sdg.esa.int/.

exploration activities are supporting sustainable development, be it for health and telemedicine, water recycling or nutrition. Technologies can be transferred on earth like for energy production, robotics, fast-deployment antennas...

Supporting sustainable development, and in particular in countries of migration for economic reasons, will help economic development of those countries and preventing part of the migration by some of the reasons for it.

It is not meant to be an ESA-only catalogue but should contribute to the elaboration of a wider tool showing also what Member States and service providers are proposing to increase the number and variety of projects proposed to the users and form the seed of a global coordination platform.

3.3.3 ESA and the UN

At the time when ESA elaborated its catalogue and presented the concept, discussions started on that matter with the UN Office of Outer Space Affairs (OOSA) following the action given by the ESA Council in June 2016. ESA is since years a strong supporter of OOSA and of UNISPACE+50 (held in June 2018), marking the anniversary of the first UN Space Conference that took place in 1968. The delegates meeting at UNISPACE+50 in Vienna have endorsed a political resolution entitled « Space as a driver of sustainable development ». ESA will continue to provide support to the UN space community and OOSA, in particular on the main challenges regarding implementations of the recommendations of the UNISPACE+50 Resolution.

Indeed, the ESA resolution entitled "Towards Space 4.0 for a United Space in Europe" adopted during the ESA Council at Ministerial level in December 2016 has several sections related to UNISPACE+50.

There is a recognition that space, and thus space agencies such as ESA, have a strong role to play in the implementation of the three global agendas:

- United Nations Sustainable Development Goals
- Paris Agreement on Climate Change
- Sendai Framework on the reduction of risks related to disasters.

All three are strongly related to sustainability and development, as well as environmental sustainability in particular.

Among the seven Thematic Priorities (TP) that were developed in the frame of the preparation of UNISPACE+50, four are more targeting sustainable development and ESA staff are involved in the relevant working groups: TP1 (Global partnership in space exploration and innovation), TP5 (Strengthened space cooperation for global health), TP6 (International cooperation towards low-emission and resilient societies) and TP7 (Capacity-building for the twenty-first century) (Fig. 5).

OOSA has elaborated the concepts of a Space Development Profile (evaluating the actual capabilities of a State) and of a Space Solutions Compendium (showing the means to increase the profile) that were also presented towards the end of 2016. ESA and OOSA agreed that the concept of the Compendium was quite close to the concept of the ESA catalogue, so the two organisations have decided to explore the

(a) Global partnership in space exploration and innovation (thematic priority 1);

(b) Legal regime of outer space and global space governance: current and future perspectives (thematic priority 2);

(c) Enhanced information exchange on space objects and events (thematic priority 3);

(d) International framework for space weather services (thematic priority 4);

(e) Strengthened space cooperation for global health (thematic priority 5);

(f) International cooperation towards low-emission and resilient societies (thematic priority 6);

(g) Capacity-building for the twenty-first century (thematic priority 7).

Fig. 5 The seven Thematic Priorities of UNOOSA (A/AC.105/1166 "The "Space2030" agenda and the global governance of outer space activities", UNOOSA, 13 December 2017, page 6)

possibility that the elements of the ESA catalogue form the seed of the OOSA Compendium that will now soon be implemented.

ESA Director General and OOSA Director signed a Joint Statement in Vienna on 19 June 2018 at the occasion of UNISPACE+50, declaring *inter alia* their intent to do a joint development of the ESA Catalogue and the OOSA Compendium.

Worth to be noted also are the discussions ongoing between ESA and the World Health Organisation (WHO). WHO takes strong interest in several ESA activities, especially for their operations in Africa. ESA Director General and WHO Deputy Director General signed a joint statement of interest in May 2018 for further cooperation.

4 Space in Africa—Regional Area for Implementation

With the catalogue described in Sect. 3.3.2, ESA wishes to better support governments, organisations, agencies or companies that wish to measure Development levels or improve them. This is possible as it helps making the link between the "space" side and the "users" side, gathering, in one tool, all activities that are felt to be contributing to the monitoring or achievement of any of the 17 UN SDGs. As mentioned before, these goals are valid for all countries in the world as reaching the SDGs is a global commitment, for developed and developing countries.

The catalogue is thus also useful when addressing the regional approaches (presented in Sect. 3.2) as the challenges faced in any of those areas can usually be linked to specific SDGs. One can for example use the word "Africa" in the search and get a certain number of examples where the activity is already done in Africa or supporting Africa, since years in some cases. There is also a specific "Space for Earth" page (as presented in Sect. 2.2) devoted to news concerning Africa.[8]

The 2063 Agenda of the African Union (AU),[9] issued in 2015, sets a strategic framework for the socio-economic transformation of the African continent for the next 50 years by building on the previous continental initiatives for growth and sustainable development.

It identifies 7 African Aspirations:

- A Prosperous Africa, based on inclusive growth and sustainable development
- An integrated continent, politically united, based on the ideals of Pan Africanism and the vision of Africa's Renaissance
- An Africa of good governance, democracy, respect for human rights, justice and the rule of law
- A Peaceful and Secure Africa
- Africa with a strong cultural identity, common heritage, values and ethics
- An Africa whose development is people driven, relying on the potential offered by people, especially its women and youth and caring for children
- An Africa as a strong, united, resilient and influential global player and partner.

Space projects and activities can clearly support the first aspiration "A Prosperous Africa, based on inclusive growth and sustainable development" and contribute to the other aspirations.

European and African States have a long tradition of exchanges and collaborations over the years, with a clear focus of the development assistance of the European States towards Africa, the closest and most historically linked continent to Europe.

In parallel to the implementation of this agenda of the AU, several ESA Member States have expressed the wish that ESA starts a reflection with them on how space could better support Africa. In particular, they are interested that ESA would help them supporting the national policies for development via the development agencies of the Member States and the many other actors like African entities or International Organisations. To that aim, and to take benefit from the work performed with the catalogue for the UN SDGs, this catalogue is used to identify the existing space activities relevant for Africa as a first step to building the initiative. This will include the activities already done in Africa and that can be further expanded or activities done outside that can be implemented also in Africa. In parallel, one will identify the relevant actors.

The following steps will cover the identification of the user needs not yet satisfied and the subsequent feasibility analysis of future projects, in the frame of the preparation of the next decisions by Member States for new programmes.

---

[8]https://www.esa.int/Our_Activities/Preparing_for_the_Future/Space_for_Earth/Africa.
[9]https://au.int/en/agenda2063.

ESA is committed to continue carrying out a multitude of activities through its programmes in support of Sustainable Development in Africa ranging from Research and Development projects and capacity building and education actions to EO data dissemination and exploitation.

This will be done with Member States that will contribute to the establishment of priorities, the identification of areas and topics to be considered, the identification of existing relevant activities, the identification of services/organisations in charge of implementation, and the next steps to be pursued.

Emphasis might be put on some specific themes that have an important relevance for Africa and in particular for selected partner countries of ESA, primarily for its economic development, and for the prevention of migration for economic reasons, and where space can highly contribute, such as education and capacity-building, water & food, health, biodiversity and climate, energy, transportation and many others.

5 The Way-Forward

5.1 Expansion of the Catalogue and Identification of "Users"

With the adoption of the United Nations 17 Sustainable Development Goals in September 2015, to be reached by all countries by 2030, the nations of the world have decided on ambitious goals for their people, prosperity, the planet, peace and partnerships. And "space" as a tool can support the measurement of and the improvement of each of these goals, already today with actual programmes, and much more tomorrow with all the programmes that are being prepared. Being financed mostly with public money implies *inter alia* that we make the most efficient use of this funding, maximising the benefits and the understanding of the benefits all can get from the space programmes. Linking them to the SDGs is a way to help "users" getting the most out of them.

And nearly each programme or activity can be linked to one or more SDGs, some in a very direct way (e.g. telemedicine for SDG3), others in a less direct way, still relevant (e.g. the results of Science missions contributing to SDG4 or SDG5). The first step is thus to make sure all relevant programmes and activities are linked, through the catalogue, to the appropriate SDGs and also associated to the regional areas specifically targeted (e.g. Africa). This will allow having test cases available when approaching potential implementing organisations.

With the support of ESA Member States, there will be also an identification of those services or organisations that should be contacted/informed for promoting space activities and making sure those that can benefit from space programmes will have the necessary relevant information.

As a next step, we will also work on the identification of gaps in services or applications that would need to be filled, with the elaboration of a proposal, or elements of a proposal, for the next decisions to be taken by ESA Member States and for an increased support of ESA to the achievement of the SDGs.

5.2 Socio-economic Benefits

The information present today in the catalogue is still not complete as socio-economic elements—quantitative or qualitative—that could help potential users or funding organisations to better understand the type and level of benefits they can expect from space programmes, applications and services, is not yet included.

A review of such elements is being performed with the support of all Directorates at ESA, as a certain number of studies are already available, and the relevant information will be included in the catalogue to support decision-making by non-space entities. This can be rather complex to define those specific indicators given the very large variety of activities concerned (programmes, applications, services, technologies…) and the difficulty in some cases to separate the effect of using space from potential other effects. But we believe it would be useful and even indispensable for adoption. A review is also performed of the SDG targets so assess which ones could be used to have an indication of the effect of space on the improvement of a target (e.g. by implementing a space application in one village and not in the next one).

5.3 Expansion of the Catalogue to Activities of Other Entities

This catalogue corresponds to a potential first step of the concept of a "Coordination Platform" that was presented already to ESA Member States, in which all agencies, organisations or service providers will be able to inform on their activities supporting sustainable development. The catalogue today only includes ESA managed projects, or commercial projects that were initiated by ESA, but it is already open to other contributors, starting with our Member States. Many, if not all, of ESA Member States, through the space agencies or government entities, or through implementations entities, have also projects to propose that complement the ESA ones. All are welcome to join the initiative in order to start building at least the European catalogue and propose the widest possible offer to the non-space world.

One Health and Space Technology—Application of Open Community Approach

Jörg Rapp, Melanie Platz and Engelbert Niehaus

Abstract

"'One Health' is an approach to designing and implementing programmes, policies, legislation and research in which multiple sectors communicate and work together to achieve better public health outcomes." (World Health Organization (2017). One Health, http://www.who.int/features/qa/one-health/en/, retreived 17 July 2018). One health and space technology are linked in terms of satellite navigation, remote sensing and satellite communication. Global navigation systems allow users to detect their current geolocation. The geolocation can be used to provide tailored information to the user via mobile devices. Satellite communication can be used among others for tele-health applications and health service delivery in remote areas. Open Educational Resources (OER) and educational justice contribute to capacity building. These OER can be tailored to the current geolocation of a user and to the profile of e.g. staff members and health services providers, especially in rural areas.

J. Rapp (✉) · M. Platz · E. Niehaus
University Koblenz-Landau—Campus Landau, Institute for Mathematics,
Landau, Germany
e-mail: rapp@uni-landau.de

M. Platz
e-mail: platz@mathematik.uni-siegen.de

E. Niehaus
e-mail: niehaus@uni-landau.de

© Springer Nature Switzerland AG 2019
A. Froehlich (ed.), *Embedding Space in African Society*, Southern Space Studies,
https://doi.org/10.1007/978-3-030-06040-4_14

1   Risk Literacy, Education and Access to Health Services

The economic pressure in the health system leads to the fact that health service delivery must be provided more and more efficiently with cost-reduction in human and medical resources. The growing gap between rich and poor leads to increasing costs and decreasing available financial resources, whereby low-income families are excluded from access to health services.[1] Risk literacy as a starting point for prevention of health risks and access to educational resources and educational equity are required to use the knowledge for local and regional risk mitigation strategies. If the gap between rich and poor continues to widen, an income inequality also implies an exclusion from certain jobs due to limited access to educational resources. In the age of digitization a growing number of such educational resources are tied in particular to digital media, software and digital education services.[2] Due to cultural, social and economic regional constraints, space technology can be used to tailor required educational resources to the local and regional requirements and constraints. This includes the availability of resources and the knowledge about local and regional health risks. Global Satellite Navigation e.g. via GPS sensors allows the implementation of regional and local information filters. To use the benefits of space technology in the health context, OER are required, that enable communities to edit and to add existing OER to their local requirements and constraints. The "One fits all"-approach does not work in general. The OpenSource area has shown that open license terms with the right of modification can lead to a reduction in development costs. This applies for example in the health context by saving 90% of the development costs through the use of an existing OpenSource infrastructure, that covers most of the features. The investment of only 10% might be necessary for the adaption of the health service delivery IT infrastructure due to local requirements.[3]

2   Open Educational Resources (OER) in Space Technology and One Health

Open educational resources (OER) are freely accessible, openly licensed text, media, and other digital assets that are useful for teaching, learning, and assessing as well as for research purposes. It is the leading trend in distance education and in the open and distance learning domain as a consequence of the openness movement.[4]

---

[1]McIntyre, D., Thiede, M., Dahlgren, G., & Whitehead, M. (2006). What are the economic consequences for households of illness and of paying for health care in low-and middle-income country contexts?. *Social science & medicine*, 62(4), 858–865.
[2]Greenhow, C., Robelia, B., & Hughes, J. E. (2009). Learning, teaching, and scholarship in a digital age: Web 2.0 and classroom research: What path should we take now?. *Educational researcher*, 38(4), 246–259.
[3]Lerner, J., & Tirole, J. (2002). Some simple economics of open source. *The journal of industrial economics*, 50(2), 197–234.
[4]Bozkurt, A., Akgun-Ozbek, E., Onrat-Yilmazer, S., Erdogdu, E., Ucar, H., Guler, E., Sezgin, S., Karadeniz, A., Sen, N., Goksel-Canbek, N., Dincer, G. D., Ari, S., & Aydin, C. H. (2015) Trends

The term OER was firstly coined at UNESCO's 2002 Forum on Open Courseware.[5] The OER concept was applied in this context to combine interconnected domains of Public Health/Global Health, Animal Health, Environmental Health and Space Technology.

According to the definition of the Open Community Approach (AT6FUI)[6] OER are a key element of capacity building by using OpenSource software if software is needed in the learning environment

- using Open Data if data is needed for learning analytic skills
- using Open Content to create Open Educational Resources
- using Open Proposal Management (see AT6FUI OpenProposal Management[7]).

One Health recognizes three health areas as interconnected[8]:

- Public Health/Global Health: the health of people and communities,
- Animal Health: the health of animals (e.g. zoonosis) and
- Environmental Health: the health of the environment and ecosystems.

For example, vultures have a certain role in an ecosystem. The health of the vulture population may have an impact on the health of other species or on public health of people living in that area. A One Health incorporates the domain of zoonoses as an important domain, in which infectious diseases are considered that can be naturally transmitted between animals and humans.

Together with space technology, One Health represents cross-disciplinary interconnected knowledge and a learning domain. The free use and re-purposing of content by others is in the interconnected setting of different domains relevant to transform educational resources and to embed these resources in a new domain, e.g. from a new scientific result and a tutorial for a new technology feature to process satellite images into the application of the knowledge in the One Health domain.

Health risks are identified in a collaborative, multisectoral, and trans-disciplinary way. Detection of environmental parameters via satellite technology or tracking of migratory birds to identify a connectivity of habitats (bird flu, H5N1) are examples that show the link between the application of space technology in the domain of

in Distance Education Research: A Content Analysis of Journals 2009–2013, International Review of Research in Open and Distributed Learning, volume 16.1 pp.330–363. https://www.academia.edu/11056576/Trends_in_Distance_Education_Research_A_Content_Analysis_of_Journals_2009-2013.

[5]UNESCO, I. F. (2002). Forum on the impact of open courseware for higher education in developing countries. *Final report.*

[6]Open Community Approach—Definition of Action Team 6 Follow-Up initiative/Expert Focus Group for Space and Global Health (2012). Retrieved from http://at6fui.weebly.com/open-community-approach.html.

[7]Open Proposal Management in AT6FUI/EFG-SGH (2015)—http://at6fui.weebly.com/open-proposal-management.html.

[8]Beasley, V. (2009). 'One toxicology', 'ecosystem health' and 'one health'. *Vet Ital, 45,* 97–110.

One Health.[9] In general, space technology is 'one tool among others'. If the application of space technology is driven by problem solving requirements, the One Health approach operates on a local, regional, national, and global level. Satellite technology can support access to spatial environmental data even on a regional level with reduced financial costs (e.g. Sentinell Programme ESA and Open Source support tools). The goal of achieving optimal health outcomes has to recognize the interconnection between people, animals, plants, and their shared environment.

If we apply the OpenSource approach in software development to the educational content and capacity building, then OER are the equivalent licensing model.[10]

The space technology community and health community do not have a long history of linkage. First applications of space technology in the field of public health appeared in the 1960s, that included tele-epidemiology and tele-medicine.[11] OER and capacity building materials from the space technology domain can be adapted for tele-epidemiology, e.g. to detect precipitation and other weather conditions that trigger mosquito abundance.[12] The capacity building material for accessing and processing satellite images combine in general OpenSource software in a software repository like GitHub[13] and the processing of satellite images e.g. for precipitation maps. This can be expanded into the epidemiological context, so that the health community is able to use the satellite images to generate risk maps e.g. for mosquito-borne diseases. It is important to use existing educational content without financial hurdles and to adapt it to local requirements and regional conditions and especially to the application context in the health domain. Space technology is in general not part of the response options in the health domain. Application of health service delivery includes capacity building for staff members working in the health domain. Also risk exposed people can have access to health service support and can e.g. be guided via applications on mobile devices.[14] The GPS is used for local health service support for patients and people living in rural communities. Health service support and capacity building differs in terms of time constraints for response. Both approaches share the generic geospatial concept that capacity building and health service support take local conditions and the interaction of learners with their environment and service providers with patients or risk exposed people into account. It is necessary to link more fields than only space

---

[9]Sader, S. A., Powell, G. V., & Rappole, J. H. (1991). Migratory bird habitat monitoring through remote sensing. *International Journal of Remote Sensing, 12*(3), 363–372.

[10]Atkins, D. E., Brown, J. S., & Hammond, A. L. (2007). *A review of the open educational resources (OER) movement: Achievements, challenges, and new opportunities* (pp. 1–84). Mountain View: Creative common.

[11]World Health Organization. (2010). *Telemedicine: opportunities and developments in member states. Report on the second global survey on eHealth.* World Health Organization.

[12]Rogers, D. J., Randolph, S. E., Snow, R. W., & Hay, S. I. (2002). Satellite imagery in the study and forecast of malaria. *Nature, 415*(6872), 710.

[13]GitHub. https://github.com/, retrieved 17 June 2018.

[14]Liu, C., Zhu, Q., Holroyd, K. A., & Seng, E. K. (2011). Status and trends of mobile-health applications for iOS devices: A developer's perspective. *Journal of Systems and Software, 84*(11), 2022–2033.

technology and the health domain to lead to a positive impact. Animal health and environmental health are linked with human health and this learning topic is linked to a geographic location. At the same time this topic is subject of learning and capacity building and respectively the improvement of risk literacy in general. There are more than one transferable subject-related contents that bridge the gap between different disciplines including local, regional and indigenous knowledge about the considered subjects. The OER approach allows other communities to transfer existing approaches to other geolocations and to adapt the learning content. On the other hand, there are naturally local and regional peculiarities that make it necessary to adapt the open learning resources to the particular out-of-school learning location.[15]

The strength of this development concept is related to the fact that the further development generates costs and/or work load. Reuse of existing OER and OpenSource software applications for a current One Health project could build on a large portion of previous development, that is free of charge. Ideally, this OER can be used directly in a teaching-learning situation without any adjustment effort.[16] In addition, educational institutions (e.g. universities, in-service training institutions, summer schools) can provide the learning resources for each individual learner tailored to the prerequisites of the learner (e.g. by creating a Wikibook for GIS and Health in 15 min[17]), but regardless of the number of students or the number of pupils, the adaptation to the concrete learning prerequisites or the respective extracurricular learning location has to be worked out only once to be valid for a certain time period.

3 Role of the Universities

Universities have a key role in accessing knowledge and for the production of scientific results in the area of One Health and in the area of space technology. The results are published in scientific papers and at the same time they are responsible for the education of students and in terms of teacher education for the integration of new scientific results in the educational curriculum.[18] This connects space

---

[15]Wolfenden, F., Buckler, A., & Keraro, F. (2012). OER Adaptation and Reuse across Cultural Contexts in Sub Saharan Africa: Lessons from TESSA (Teacher Education in Sub Saharan Africa). *Journal of Interactive Media in Education.*

[16]Pirkkalainen, H., Thalmann, S., Pawlowski, J., Bick, M., Holtkamp, P., & Ha, K. H. (2010, June). Internationalization processes for open educational resources. In *Workshop on Competencies for the Globalization of Information Systems in Knowledge-Intensive Settings, ICSOB.*

[17]Risk Management/Tailored Wikibooks. (2018, March 5). Wikiversity. Retrieved 08:07, March 5, 2018 from https://en.wikiversity.org/w/index.php?title=Risk_Management/Tailored_Wikibooks&oldid=1826994.

[18]Zeichner, K. (2006). Reflections of a university-based teacher educator on the future of college-and university-based teacher education. *Journal of teacher education, 57*(3), 326–340.

technology and One Health with teacher education as multipliers for education with OER. OER produced by universities for capacity building and learning along with new scientific results can be integrated e.g. in Wikiversity underlearning topics on a more frequent base (see Risk Management Course in Wikiversity[19]). Teachers need a digital literacy for using, adapting, creating and discussing OER,[20] so that the results can be adapted to local and regional or social constraints of the groups of learners and to the geolocation where the learners focus on.

3.1 Multiplier Training

If free access to educational resources is a goal for universities, OpenSource software and OER become the bundle for a free and open learning environment.[21] The teachers have to be prepared for the later handling of OER and the software products, that are needed in the learning environment. Ubiquitous mobile devices provide the IT infrastructure for learning and capacity building. Sharing of content is required via Bluetooth or Wireless Local Area Networks, due to the fact that mobile internet connectivity might not be available in rural areas of health affected countries. In order to replicate a successful OER documented case study, the Open Community Approach (OCA) integrates OpenSource and Open Content to minimize the replication costs for the infrastructure.[22] Furthermore, the capacity building material applies the software independent content in the space technology domain and One Health with user-oriented training that refers to the region and the available resources. Even if commercial resources are available at one community, the application of the approach might be not transferable because another community does not have the financial resources for the commercial product. The OpenSource software can be used for example for the evaluation of data in the professional context, for the substantiation of technical conclusions about risk mitigation options and for visualisations e.g. of risk maps with OpenSource products. The multiplier training of the teachers at universities should lead to the fact that they expand their knowledge about commercial products to training with OpenSource software as well, so that teachers have expertise in the integration of OpenSource products into the educational process about space technology and global health. If the teachers are not exposed to OER and OpenSource in their own education, it is difficult to perform qualified decision making between open and commercial approaches and to integrate the open educational contents and free

---

[19]Risk Management. (2018, June 18). *Wikiversity. Retrieved 17:52, June 18, 2018 from* https://en.wikiversity.org/w/index.php?title=Risk_Management&oldid=1889264.
[20]Pianfetti, E. S. (2001). Focus on research: Teachers and technology: Digital literacy through professional development. *Language Arts, 78*(3), 255–262.
[21]Caswell, T., Henson, S., Jensen, M., & Wiley, D. (2008). Open content and open educational resources: Enabling universal education. *The International Review of Research in Open and Distributed Learning, 9*(1).
[22]Open Community Approach. (2018, May 13). *Wikiversity. Retrieved 15:59, May 13, 2018 from* https://en.wikiversity.org/w/index.php?title=Open_Community_Approach&oldid=1865926.

OpenSource software into the learning processes for linking space technology and One Health.

3.2 Reproducibility and Reproducible Science

Overall, OERs and OpenSource software are subcomponents of an educational system that supports reproducibility and verifiability of scientific results. Free access to software is therefore not an end in itself but reduces the financial barriers to public participation in validation mechanisms. If this educational content is open, it can be used beyond the university context, it can be also made freely accessible to the public at a certain geolocation for further education and training of citizens for validation of facts and insights in consequences for the current geolocation. Global Satellite Navigation Systems (GSNS) allow the allocation of educational content to a certain geolocation e.g. by the application of web-based tools to assign digital content to Open Street Map.[23]

3.3 Participation in Research Results

The approach of Citizen Sciences allows the report of health issues by citizens.[24] Data collection can be performed by software tools like Open Data Kit (ODK).[25] An app installed on a mobile device can add the geolocation of the collected data. E.g. observations of migratory birds can be documented by citizens and the aggregated data can be used for cross-validation of tracking of migratory birds with GPS tracking devices (see Movebank: a global database for animal movement Wikeski MPI Konstanz[26]). OER can be the citizen science approach itself by the application of data collection tutorials with ODK, comprehension of scientific methods of tracking and analysis of animal movements.

4 Wikiversity and OER

Wikiversity is a collaborative repository of OER for[27]

---

[23]OpenStreetMap and its use as open data (2018, July 10). Retrieved from https://www.e-education.psu.edu/geog585/node/738.

[24]Den Broeder, L., Devilee, J., Van Oers, H., Schuit, A. J., & Wagemakers, A. (2016). Citizen Science for public health. *Health promotion international*, *33*(3), 505–514.

[25]Open Data Kit (2018, July 10). Retrieved from https://opendatakit.org/.

[26]Movebank: a global database for animal movement. (2018, July 10). Retrieved from https://www.orn.mpg.de/33343/Movebank.

[27]Wikiversity: What is Wikiversity? (2018, July 10). Retrieved from https://en.wikiversity.org/wiki/Wikiversity:What_is_Wikiversity%3F.

- Non-formal education (e.g. for farm workers to apply vegetation indices for crop health monitoring and reduced application of agrochemicals and less exposure to chemicals in the work flow).
- Primary Education—understand spatial concepts of (invisible) risks (e.g. exposure to chemicals and toxic substances in our environment and to learn about risk mitigation strategies).
- Secondary Education—application of mobile devices in a learning environment and participation in Citizen Science projects.
- Tertiary Education—universities and academics in-service training.

Wikiversity uses the underlying MediaWiki[28] as OpenSource software, so that authors do not have to learn a new software interface but can use a well-known from Wikipedia. The learning process guides receptive users of the resources towards a more advanced role as author, editor, validator of content. Especially the validation of contents with scientific resources by Wiki community members is very important for quality assurance. Even if anonymous editing with report of the IP-addresses is possible, a transparent editing process can be performed by a login and by providing at least the affiliation of the editor.

4.1 Licensing OER

The consideration of a licensing concept (e.g. Creative Commons as applied in Wikiversity and Wikipedia) was built as open licensing within the existing framework of intellectual property rights, so that legal issues and violation of copyright can be handled. Especially commercial use can not be allowed, so that all adaptions of the work remain open to the community. Nevertheless, commercial use of Creative Commons products may be granted by the licensing model, so that companies can take Creative Commons products and use them in business settings. In general the intellectual property rights must be applied as defined by relevant international conventions. OER respects the authorship of the work and contributions to OER products.[29]

4.2 Drivers for OER

Drivers for the application of OER are to:

- maximize public access to educational resources especially in developing countries and translate them to local language.

---

[28]MediaWiki/Engine. (2016, June 25). *Wikiversity. Retrieved 02:51, June 25, 2016 from* https://en. wikiversity.org/w/index.php?title=MediaWiki/Engine&oldid=1579839.
[29]Downes, S. (2007). Models for sustainable open educational resources. *Interdisciplinary Journal of E-Learning and Learning Objects, 3*(1), 29–44.

- trigger innovation by accessing the educational resource and learn about new concepts and approaches.
- reduce developing costs authors do not have to build from scratch, but can build, remix or translate existing capacity building material.

The Open Content Development Process analyses the required resources and even if the OER and OpenSource software is free, the implementation and application may need human resources that are added to the TCO (Total Cost of Ownership) similar to commercial software and educational resources.[30]

5 Open Source Statistics

On January, 16th 2012, the aggregated number of downloads at Sourceforge,[31] an Open-Source Software Platform, was 4349930 (4.4 Mio) for 11 h.[32] Now we take the number of downloads as reference values for activities at this software repository. In the period of time the Sourceforge statistics showed

- 3963 code commits (0.1% workload), which means number of updates for all software repositories,
- 4105 forum posts (0.1% workload), which include questions, recommendations and discussions about a certain software product,
- 443 bugs tracked (0.1% workload), which is the error report of users about the software.

The number show a basic ratio between the 100% of users that just benefit by downloading free software from Sourceforge and the work as developers or contributors and supporters of the software development. Error report of tests and feedback are relevant contributions for the further development and bug fixes of the OpenSource repository. The ratio between contributors and users that just download was 2 out of 1000. On the one hand, this specific ratio is not representative for other open collaborative environments. On the other hand, it shows a basic rule of thumb for expectation for contributions from the community to a specific collaborative effort. A contributor might regard this ratio as discouraging, but the ratio can be regarded as very positive in terms of reaching people. With 2 activities the developer can cause 1000 download activities of users. This can be transferred to OER in terms of interest in open content.

---

[30]Schaffert, S. (2010). Strategic integration of open educational resources in higher education. In *Changing Cultures in Higher Education* (pp. 119-131). Springer, Berlin, Heidelberg.
[31]SourceForge: The Complete Open-Source Software Platform, https://sourceforge.net/, retrieved 17 July 2018.
[32]SourceForge Support: Download Stats API (2012, February 10). Retrieved from https://sourceforge.net/p/forge/documentation/Download%20Stats%20API/.

6   The Potentials of Wikiversity

Wikipedia is a well-known community driven knowledge base. For capacity building Wikiversity is used. The main differences are that

- Wikipedia is designed as Encyclopedia and
- Wikiversity is a repository of Learning Resources.

Both products share a common Wiki software infrastructure called MediaWiki[24]. For this paper risk management is one starting point to explore a basic risk management cycle and spatial risk management strategies as learning resources in Wikiversity.

6.1   From Wikiversity Article to Presentation

Wikiversity can be regarded as an educational sink for OER. In Wikiversity articles can be converted directly from Wikipedia into a web-based presentation, that can be commented with a stylus on a touch screen. The following screen shot (Fig. 1) shows PanDocElectron[33] as a document conversion tool for wiki articles, that is able to convert the Wikiversity source into different formats including

- RevealJS as web-based presentation,
- Word,
- LibreOffice Writer,
- LaTeX,
- and many others.

The basic principle is to maintain not only articles with learning tasks, quizzes, etc. in Wikiversity but also to generate presentations for educational settings.

6.2   Wikipedia Book Creator

The Wikipedia Book Creator is a software tool, that allows users to aggregate different articles of choice into a tailored book that fits to the heterogeneous requirements of the learning objective and prerequisites of the learner. This concept supports the tailored generation of books as learning resources according to the needs of the learners.[34] The learning video on Wikiversity shows the basic concept.

---

[33]PanDocElectron (2018, July 10). Retrieved from http://niebert.github.io/PanDocElectron/.
[34]Risk Management/Tailored Wikibooks. (2018, March 5). *Wikiversity. Retrieved 08:07, March 5, 2018 from* https://en.wikiversity.org/w/index.php?title=Risk_Management/Tailored_Wikibooks&oldid=18 26994.

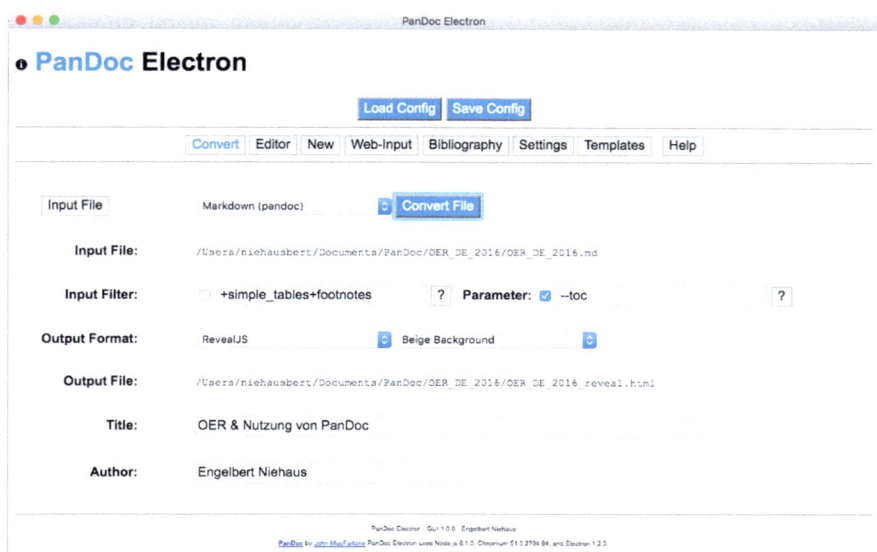

Fig. 1 Screenshot of PanDocElectron GUI

Currently (2018) the Book Creator undergoes major updates to improve the features, maintenance and performance of the tool.

6.3 Ecotoxicology, Space Technology and One Health

As mentioned above, space technology and remote sensing is able to detect environmental parameters (e.g. land coverage[35] or harmful algae bloom[36]). Especially on larger scales space technology supports the detection of harmful algae bloom, study the spatial distribution over time and create One Health warnings for the local population. The toxicity of these algae to humans is due to the presence of algal toxins.[37] The temporal change of algae bloom can be studied along with climate change issues and its impacts on community, ecosystem, and biosphere levels. Economic impact arises from the impact and awareness in the local fishery and aquaculture.

---

[35]Thompson, M. (1996). A standard land-cover classification scheme for remote-sensing applications in South Africa. *South African Journal of Science*, *92*(1), 34–42.

[36]Tang, D. L., Kawamura, H., Doan-Nhu, H., & Takahashi, W. (2004). Remote sensing oceanography of a harmful algal bloom off the coast of southeastern Vietnam. *Journal of geophysical research: oceans*, *109*(C3).

[37]Van Dolah, F. M. (2000). Marine algal toxins: origins, health effects, and their increased occurrence. *Environmental health perspectives*, *108*(Suppl 1), 133.

Ecotoxicology and public health are interconnected in the context of precision farming[38] or vector control with insecticides.[39] The exposure of agrochemicals to farm workers and to the environment can be studied in a multidisciplinary field, which integrates toxicology, ecology, public health and space technology. The underlying principle is to minimize the exposure of the environment and humans to chemicals by preserving the harvest yield. Precision agriculture is related to remote sensing, whereby the crop health status is detected via the Normalized Difference Vegetation Index (NDVI).[40] The primary economic objective is to use less agrochemicals for the same harvest yield. The benefits in the long run in the economic setting are the systemic link of biodiversity and pollination as an ecosystem service for agriculture. Long term public health benefits can be accomplished in the context of nephrotoxic impacts of agrochemicals on human health.[41]

6.4 Reader to Author

Working at the intersection between space technology and One Health the learning process for the reader identifies in general missing gaps in the learning resource that can be filled by the knowledge of the reader/learner. Especially when a learning resource was designed first by an author from the space technology domain, the content might be transformed to make it comprehensive for learners from the health domain and vice versa. The learning concept from "reader to author"[42] encourages learners to become active contributors to Wikiversity, which can be regarded as a learning task for advanced Wikiversity authors in general. In this sense, Wikiversity authors can visit the exploratory study on OpenSource education and improve the content in terms of

- technical skills: to be able to edit articles, correct typos, validate citations, improve the citations, so that the references are processed in MediaWiki standards and add links from the One Health domain to space technology and vise-versa in Wikiversity.
- semantic improvement: analyze the link between the presented topic space technology and One Health and add value to OER in general.

---

[38]Candiago, S., Remondino, F., De Giglio, M., Dubbini, M., & Gattelli, M. (2015). Evaluating multispectral images and vegetation indices for precision farming applications from UAV images. *Remote Sensing*, 7(4), 4026–4047.

[39]Barnes, K. I., Durrheim, D. N., Little, F., Jackson, A., Mehta, U., Allen, E., ... & White, N. J. (2005). Effect of artemether-lumefantrine policy and improved vector control on malaria burden in KwaZulu–Natal, South Africa. *PLoS medicine*, 2(11), e330.

[40]Calculating a Regional NDVI. (2018, July 14). *Wikiversity. Retrieved 02:41, July 14, 2018 from* https://en.wikiversity.org/w/index.php?title=Calculating_a_Regional_NDVI&oldid=1897552.

[41]Wikipedia contributors. (2018, June 12). Mesoamerican nephropathy. In *Wikipedia, The Free Encyclopedia*. Retrieved 11:38, July 16, 2018, from https://en.wikipedia.org/w/index.php?title=Mesoamerican_nephropathy&oldid=845532163.

[42]Reader2Author. (2018, January 21). *Wikiversity. Retrieved 00:26, January 21, 2018 from* https://en.wikiversity.org/w/index.php?title=Reader2Author&oldid=1810935.

- create submodules: if applicable create submodules of OER in the health domain in Wikiversity with additional information on space technology, so that learners with a health background are able to use approaches from the space technology domain to complement classical approaches in the health domain.
- watchlist: put the article you work on to your own watchlist as authenticated Wikiversity user, to keep informed on changes about the topic. The "blue star" in the top menu indicates for the current article, if the article is on the author's watchlist. A reader puts himself on the watchlist to be able to act as a reviewer whenever the content changes in that article.

6.5 WikiJournal

Wiki journals are used in the context of OER as capacity building and learning resource, which is open access and does not charge an author for publication. At the same time a peer-reviewing process is implemented.[43] In comparison to the standard philosophy of peer-reviewed journals, the Wiki approach regards a reviewer as a community member that contributes to the quality of a paper. Transparency of the reviewer's background show the contributor's potentials and benefits especially in an interdisciplinary setting. A reviewer from space technology could add valuable comments to a public health paper in terms of tele-epidemiology and the application of space technology, but this WikiJournal reviewer might not have asked by Journal Editors in the Health domain, because they were not aware of these new developments in the space technology domain due to their academic background. Rejection of authors to pickup these cross domain reviewer comments may trigger other researchers and developers to pickup certain ideas and to build on this reviewing discussion between space technology and health issues. So, a transparent open reviewing process is a learning resource itself and Wikiversity can be regarded as an open innovation ecosystem for the intersection of space technology and health. Finally, the versioning system for a scientific paper is able to show the development of the paper (who, did when what?). If scientists do not want to show the development process of the paper, it is possible to submit an article as a whole e.g. for the sake of clarity and comprehensibility, to hide the contribution story or to acknowledge certain contributions with a few lines at the end of the paper.

---

[43]WikiJournal of Medicine. (2017, December 10). *Wikiversity. Retrieved 17:20, December 10, 2017 from* https://en.wikiversity.org/w/index.php?title=WikiJournal_dof_Meicine&oldid=1788945.

7 Dynamic Spatiotemporal Documents Explained on the Example of Spatiotemporal Patterns

Space technology provides remote sensed data and the health domain processes them. Spatial applications of risk mitigation strategies are e.g. compared with the prevalence of non-traditional Chronic Kidney Disease (ntCKD) in agriculture. Dynamic spatiotemporal patterns are scientifically analyzed and the papers are generated based on collected data. In comparison to classical approaches of publication in a paper, the dynamic spatiotemporal generation of documents incorporates meta layers of "What if" logic and takes a permanent inflow of monitoring data and remote sensed data as input value of a dynamic repeated analysis of the data. The KnitR-package in R[44] allows the implementation of dynamic spatiotemporal analysis into the OpenSource software R for statistics (see https://www.rstudio.com/).[45] The underlying principle uses R-code for analysis, to retrieve spatial data and instead of copying the generated diagrams into an office document, an R-Markdown document is generated with inserted R-code-chunk that generate e.g. a diagram on the fly from the retrieved data. A scientific expertise of the "What if" logic of analysis is transparently encoded in the R-Code. The generated document is not a classical static paper because it incorporates updated data. New versions of the results integrated in Wikiversity journals show the version history of the paper and a reference of the R-Markdown paper on a Git-Repository allows other scientist to reproduce the results. The underlying technical concept combines remote sensed data and the data analysis of One Health into a dynamic paper generation for spatiotemporal disease patterns, that can be used as a spatial decision support system in a dynamic scientific publication. Furthermore, any use of R-code in R-Markdown as code chunks can be used in the web-based application R Shiny[46] for visualizing the results. Those scientists who copied and pasted diagrams from the statistics software R into an office document in a time consuming way can use their R-Code for analysis as code-chunks in R-Markdown and wrap the text of the article around the code chunks. The text of the article can be dependent on the results of statistical analysis. From R-Markdown different formats of office documents can be generated:

- R-Markdown to LibreOffice/Word
- R-Markdown to PDF and LaTeX and
- several other formats due to the application of the open source tool PanDoc,[47] developed by John MacFarlane, as a "swiss-army knife" for document conversion.

---

[44]KnitR. (2018, April 1). *Wikiversity. Retrieved 05:58, April 1, 2018 from* https://en.wikiversity.org/w/index.php?title=KnitR&oldid=1841817.

[45]Rstudio (2018, July 10). Retrieved from https://www.rstudio.com/.

[46]Shiny from Rstudio: Interact. Analyze. Communicate (2018, July 10). Retrieved from https://shiny.rstudio.com/.

[47]PanDoc. https://pandoc.org/, retreived 17 July 2018.

8   Summary and Conclusion

The paper documents the link between Open Educational Resources for space
technology and One Health e.g. in Wikiversity as a repository for OER. Wikiversity
is a product of the Wiki Foundation that runs Wikipedia as well. In this sense the
foundation is a main driver for preserving open access community knowledge. The
following items are key elements of the paper

- Educational Content Sink: learn about Wikiversity as Educational Content Sink
  that supports the implementation of links between space technology and One
  Health in an open environment.
- Technical Support OER: technical analysis and dynamic document generation
  with markdown languages e.g. Wikiversity and in R with the KnitR package.
  Other document formats can be generated by the application of PanDoc that is
  integrated in the KnitR approach as well.
- Generate presentation slides from Wikiversity: convert Wikiversity content
  about One Health and space technology into slides for presentations. Slides can
  be maintained jointly by the space technology community and the experts in the
  One Health domain. Selected articles of choice can be generated with
  PanDocElectron into a web-based presentation.
- OER and mobile devices: the spatial analysis and dynamic generation of
  web-based content with statistic software can be used to create tailored support
  for health risk exposed citizens by using the global positioning information from
  mobile devices to tailor the information for the owner of the mobile device. To
  accomplish this objective, it is necessary to create a strategy document how to
  use, share and adapt OER for mobile devices and how to create an IT infrastructure that tailors the available information according to requirements and
  constraints of users in a transparent way. This is necessary that users can trust
  the results and scientist are able to reproduce the filter mechanisms in the OER
  repository of knowledge base.
- OER and risk literacy: the objective of the OER application is to improve risk
  literacy and to derive rationals for the application of OER in the context of risk
  mitigation strategies for space technology and One Health, especially in
  developing countries. The infrastructure should be easy to access and the
  knowledge base must have a licensing model that allows the adaption to local
  and regional requirements and constraints, which include the right to translate
  the OER into local and regional languages. Open Badge Management[48] can be

---

[48]Open Badges. (2017, November 27). *Wikiversity. Retrieved 11:41, November 27, 2017 from*
https://en.wikiversity.org/w/index.php?title=Open_Badges&oldid=1784927.

used as digital certification and incentive of a learning process. Meta data in the Open Badges show further information about the course on space technology and/or One Health topics.[49]

- Localization of OER: tailor educational resources to specific requirements and constraints of the target group or use-case.
- Digital Signature: the concept of digital signature allows that an expert (e.g. from WHO) could sign an OER document with a private key. This adds a digital signature to the digital document. If the digital document is e.g. an office document (LibreOffice Writer) and anybody changes only one single character in the source document, then the digital signature is invalid because it is not provided in the original status e.g. by WHO in the quality assured version. Nevertheless, the OER approach allows the translation of the source document in a local language or with alterations according to the application of space technology in the health domain. Furthermore, the authors who changed or translated the signed document by WHO can validate that the WHO document was not altered by someone else with the application of a public key provided by a quality assuring institution for the issued document.
- Optimization: learning resources can be optimized in a community driven approach by the analysis of the different use-cases of the learning resources.

Finally, the whole proposed application focuses on risk assessment and response according to risk in the One Health domain and on an open capacity building and learning concept to leverage the potential of space technology in the health domain. It complements the classical approach in the human, animal and environmental health domain with additional analytic resources and a dynamic document management and versioning system for analysis and visualization.

Author Biographies

Jörg Rapp has studied environmental sciences and is a Ph.D. candidate and research associate in the working group for Mathematical Modeling at the Institute of Mathematics of the University of Koblenz-Landau, Campus Landau. His research focuses on the interface between applied mathematics and environmental science.

Dr. Melanie Platz Since 2011 Dr. Melanie Platz works as research associate in the working group for Mathematical Modeling at the Institute of Mathematics of the University Koblenz-Landau, Campus Landau. Her current research focus is on the implementation of electronic systems for performance verification in the university sector as well as the use of electronic devices in school education. She is currently visiting professor in the working group Didactics of Mathematics at the University of Siegen.

---

[49]E. Niehaus, M. Platz, A. Botha, M. Herselman. (2017). Using Digital Badges in South Africa informing the validation of a multi-channel Open Badge system at a German University, in IST-Africa 2017 Conference Proceedings, Paul Cunningham and Miriam Cunningham (Eds), IIMC International Information, Management Corporation, 2017.

Prof. Dr. Engelbert Niehaus is head of the working group for Mathematical Modeling at the Institute of Mathematics and head of the Computer Science Centre of the University Koblenz-Landau, Campus Landau. His research interests include the use of open educational resources in educational processes as well as the development of mathematical methods that can be used in the area of risk mitigation. He is a member of the United Nations OOSA Expert Focus Group Expert Committee on Space and Global Health, where he is responsible for mathematical modeling in the development of an "epidemiological early warning system".

Security in Africa: A Perception of Ongoing Developments

Gerald Hainzl

Abstract

Discussing security is a complex endeavor as there is no agreed definition. Dividing it in a narrow and broader perspective may help to analyze violent conflicts and the reactions to it as well as long-term programmes for a less violent future. Instruments and organizational structures vary over time as a reaction to the different challenges. These dynamic developments and interest driven engagement are a reflection of the narrow perspective. The broader perspective with its long-term focus on human security may have a weaker short-term impact, but reduce violent conflict on the long run. International organizations and states perform different tasks in providing security in specific situations.

1 Introduction

What is "security" from the point view of a state, a group of people or individuals is tied to the understanding of how a state, a group of states or a society views threats to its or their group or individual existence. With more than fifty states in Africa with quite divers populations, several regional economic communities and other actors the answers to the question seems to differ substantially according to the security environment of the specific state or organisation. Most probably the answer will be constructed according to a given situation in a specific moment of history. How security is perceived and embedded into the state applications varies across

G. Hainzl (✉)
Institute for Peace Support and Conflict Management, Austrian National Defence Academy, Vienna, Austria
e-mail: gerald.hainzl@bmlv.gv.at; gerald.hainzl@me.com

© Springer Nature Switzerland AG 2019
A. Froehlich (ed.), *Embedding Space in African Society*, Southern Space Studies,
https://doi.org/10.1007/978-3-030-06040-4_15

the continent. As the Africa Center for Strategic Studies puts it ‚Africa's security environment is characterized by great diversity'.[1] A fast and constant changing security environment makes it necessary for all actors to adapt and change their concepts.

The vast field of security can be divided into a narrow and broader view as well. While the narrow view seems to look just at goal number 16 of the sustainable development goals (SDG's, also referred to as global goals) a broader view should also take the other goals into account.

In this article a narrow and a broad view on security will be used. A narrow view on security (often called hard security) which focusses more or less only on physical violence from actors inside the country against individuals or groups and from actors outside the country against the state and its institutions. A broad view which encompasses all aspects of human security.[2] Furthermore it is not based on an analysis of all relevant documents issued by the African Union, Regional Economic Communities and individual states, but on the perception of what is put into practice, when it comes to security issues.

2   African States and Security

The provision of security for its citizens is one of the core tasks of the state. Although there are organizational differences all over Africa (and the world) in a narrow perspective police forces, military forces for inner and external security as well as state security agencies are meant. Depending on the state of the respective African state (out of 54) the ability to provide security varies significantly. While some countries are able to fulfill all necessary tasks to ensure a certain level of security, others are not. And some are even not able to exert authority over regions where they claim sovereignty. Classifying or ranking countries along the level of ability neither is possible nor useful. It is not possible, because of a lack of reliable validated data and it is not useful as it would not have any practical consequence or just serves a purpose from a specific, often external political perspective.[3]

Discussing security in Africa from an external perspective immediately violent conflicts come to ones mind from the Democratic Republic of the Congo (DRC) to South Sudan to Northern Nigeria to Somalia. From an external perspective all these conflicts look quite similar, but are they? If, e.g., the heterogeneity of African states is a reason for conflict, why is the most homogeneous country on the continent

---

[1] www.africacenter.org/security-topics/, accessed July 27th, 2018.

[2] The author is very well aware that there are many schools of thought and much more definitions. For the sake of simplicity basic working definitions will be used. See e.g. Tadjbakhsh, Shahrbanou: In Defense of the Broad View of Human Security, In: Routledge Handbook of Human Security, Routledge 2013. On the concept of security see Baldwin, David A. The concept of Security. In: Review of International Studies. 1997. 23. 5–26.

[3] For a critical view see Williams, Paul D., Thinking about Security in Africa, International Affairs, Vol. 83, No. 6, Africa and Security (Nov. 2007), 1021–1038.

(Somalia) almost always the textbook example as a failed state? Where and how a group acts seems largely dependent on the political/economic/religious etc. narrative and the identity concept of the group itself. This may be one of the reasons why some conflicts stay within state boundaries while others are part of systems of conflict and violence, which have a negative impact on a whole region. Although there is some external influence in the eastern parts of the DRC neighboring countries are mainly effected by refugees, if at all. The numerous violent groups confine their area of operations mostly to the DRC. On other hand, Boko Haram in Northern Nigeria is operating across the borders as well. That said, there are interesting consequences not only for the state, but for the Regional Economic Communities as well. In regions where conflicts are more or less confined to the state, the pressure for groups of states to cooperate in the field of security is not as high as in regions with cross-country challenges.

The broad view on security on state level is largely dependent on the challenges the country faces. While the political commitment on international, regional or sub-regional level most of the time is given, for different reasons several countries lack concrete domestic actions, when international treaties and arrangements should be implemented.[4] The less binding these agreements are the less the pressure is to do so.

3 Regional Economic Communities and Their Impact

As mentioned above the pressure to cooperate within a Regional Economic community grows when security becomes a cross-border issue and acting alone is not possible any more. And if they challenge is even bigger new forms of cooperation are started.

Since the 1990 the slogan "African solutions to African problems" had become a central catchphrase, when security issues were discussed. Depending on who used it in which context, a large variety of political interpretations were given and numerous articles were written on this topic. While it had different meanings to different actors within and outside Africa during the last few years inner African dynamics show a gradual improvement of the willingness of African countries to engage in restoring non violent environments.[5] In the center of these dynamics are the Regional Economic Communities. A trigger of these developments, especially in West Africa, was the cooperation in the fight against terrorism.

---

[4]See e.g. on the challenges of implementing international treaties: Aliyu Ahmed-Hameed:
  The Challenges of Implementing International Treaties in Third World Countries: The Case of Maritime and Environmental Treaties Implementation in Nigeria, Journal of Law, Policy and Globalisation, Vol. 50, 2016.
[5]Recent examples are political and military interventions of ECOWAS in The Gambia (2017) or in Burkina Faso (2015).

The Regional Economic Communities (RECs) have security as one of their political goals as well. It is highly dependent on the challenges the RECs faces, on which fields of cooperation they concentrate. In general with the exception of the Union of the Arab Maghreb (UMA) all RECs try to increase cooperation.

Increased cooperation takes place not only within one REC, but also between them. A meeting of the heads of states and government between the Economic Community of West African States (ECOWAS) and the Economic Community of Central African States (ECCAS) took place in July 2018[6] aiming at strengthening the ties in order to create conditions for sustainable peace and a secured environment. Although they already held a joint meeting on security in 2013[7] and agreed a certain level of cooperation changing and increasing security challenges made it necessary to define/redefine and make cooperation more effective. As the final communiqué states the areas where increased cooperation in West and Central Africa is needed are "terrorism, human, drugs and arms trafficking, money laundering and cybercrime". Collective response is necessary as several initiative to tackle these problems involve countries from both RECs (e.g. G5 Sahel or the Multinational Joint Task Force of the Lake Chad Basin). In order to assure and strengthen the response early warning as well rapid response mechanisms are necessary, while a regional framework on the constitutional principles provide the legal means. Closer cooperation in the fields of military training and operations and law enforcement as well as intelligence sharing are part of the struggle for increased security.[8]

The role RECs can play in improving security is also shown by the Southern African Development Community (SADC). The role the regional body can play in the not so distant past was e.g. shown in Lesotho, Madagascar and the DRC,[9] although it is largely dependent on the will of the receiving country.[10]

4 New Forms of Cooperations

The dynamics in African security policy are driven by the need to react to various challenges. States try to find new ways of cooperation as "traditional" organisation seem to be either to slow in their political reaction or more than one of these would

---

[6]Joint Summit of ECOWAS and ECCAS Heads of State and Government on Peace, Security, Stability and the Fight against Terrorism and Violent Extremism, Final Communique, Lomé, 30th July 2018.

[7]ECOWAS-ECCAS Joint Summit held on 25th June 2013 in Yaoundé, Republic of Cameroon.

[8]The call to cooperate is not a very recent phenomenon. In September 2012 the newspaper Leadership (Abuja) reported: West Africa: Ecowas Reiterates Call for Closer Cooperation Among Law Enforcement Agencies.

[9]Ebrahim, Shannon, SADC summit focuses on bringing stability to Lesotho, DRC and Madagascar, April 25th, 2018, https://www.iol.co.za/news/africa/sadc-summit-focuses-on-bringing-stability-to-lesotho-drc-and-madagascar-14637285, accessed August 27th, 2018.

[10]Louw-Vaudran, Liesl, Ls SADC's work in the DRC over?, Institute for Security Studies, ISS today, August 22nd, 2018, https://issafrica.org/iss-today/is-sadcs-work-in-the-drc-over, accessed August 27th, 2018.

be involved. The Multinational Joint Task Force of the Lake Chad Basin and the G5 Sahel are examples for this approach. RECs are not necessarily in opposition to these developments. E.g. ECOWAS and ECCAS welcome the commitment of some of their member states and ask the others to provide assistance.[11]

A critical issue for most African initiatives in the field of security is funding. Requesting money from donors however can lead to situations where both sides are unhappy with the outcome. As an alternative ECOWAS ECCAS are calling on the United Nations to place it under Chap. 7 of the UN charter in order to assure multilateral sustainable funding.[12]

The Multinational Joint Task Force of the Lake Chad Basin (MNJTF) is tasked to fight Boko Haram after the groups area of operation spread from Northern Nigeria to the neighboring countries and was authorized by the Peace and Security Council of the AU in 2015.[13] Although criticized for its lack of effectiveness it continually improved and had several successes in the fight against the terrorist group.[14] The G5 Sahel[15] was created in 2014 by the heads of state of Burkina Faso, Chad, Mali, Mauritania and Niger "to co-ordinate strategies and policies on defence, security, governance, infrastructure and resilience in the Sahel."[16] When its joint forces were launches in 2017 it had the support of the UN and MINUSMA, the UN mission in Mali was authorized by a resolution of the security council to provide logistical and operational support.[17] Like in almost every African security related endeavor financing was and is a critical issue: "G5 Sahel countries have each pledged EUR 10 million; Saudi Arabia will contribute USD 100 million, the European Union EUR 50 million and France will provide equipment worth EUR 8 million. Nonetheless, there is still a significant budget shortfall and the question of how the joint force will cover its operating costs – about EUR 120 million per year by some estimates – is also still unresolved."[18] The external financial contributors may expect a return on theirp investment, but their agendas are often not so clear. In the case of the EU and France one can suspect migration and terrorism as the driving forces behind, while Saudi Arabia's interest could be to gain influence in Africa to foster its aspirations as a regional power.

---

[11]Joint Summit of ECOWAS and ECCAS Heads of State and Government on Peace, Security, Stability and the Fight against Terrorism and Violent Extremism, Final Communique, Lomé, 30th July 2018, Chapter 17.

[12]AMISOM in the first years of the mission, when funding was provided i.a. by the European Union.

[13]For more information on MNJTF and first hand information see https://www.mnjtf.org, accessed August 26th, 2018.

[14]see Assavano, William, Abatan, Jeannine Ella A and Sawadogo, Wendyam Aristide, Assessing the Multinational Joint Task Force against Boko Haram, Institute for Security Studies, West Africa Report, Issue 19, September 2016.

[15]Detailed informations on G5 Sahel can be obtain at http://www.g5sahel.org, accessed August 26th, 2018.

[16]The G5 Sahel and its Joint Force. West Africa Brief. http://www.west-africa-brief.org/content/en/g5-sahel-and-its-joint-force, accessed August 26th, 2018.

[17]Resolution 2391 (2017). Adopted by the Security Council at its 8129th meeting, on 8 December 2017.

[18]The G5 Sahel and its joint force. op. cit.

The above described new forms of intervention are only tackling issues dealing with a narrow perspective on security while they lack a broader view with a more comprehensive and lasting concept of security. One reason may be that they were created for a specific kind of violent threat, while the concept of human security is dealt with by the AU or the RECs.

5 The African Union and Security

When the African Union (AU) was founded almost two decades ago as the successor organisation to the Organization of African Unity (OAU) the continental organization transformed into an international body with a many tasks to fulfill. Peace and Security were always one of the priorities to the AU as it is reflected in the organizational structure by the Peace and Security Council (PSC). When the PSC was established in 2002 the founding document reiterates the commitment to early warning, preventive diplomacy, peace making as well as peace support operations and interventions.[19] It goes without saying that the idea of a common continental Defense and Security policy is included in the Protocol. The Constitutive Act of the AU[20], adopted in 2000, already had a focus on peace and secured included.[21] And the AU was very ambitious to put its political promises into action. AU missions were active in Burundi and in Sudan. An African Union-led Regional Task Forces was fighting the Lord's Resistance Army in Uganda, an African-led International Support Mission to Mali and an African-led International Support Mission to the Central African Republic show the commitment of the AU. The African Union/United Nations Hybrid operation in Darfur (UNAMID) was another attempt to contribute to peace and security, where a state or states failed to provide.

The biggest challenge for the AU in terms of peace and security so far seems to be Somalia. In 2007 the African Union Mission in Somalia (AMISOM) was started as an ambitious project and the to date the mission is still on the ground. At the beginning 8.000 troops were authorized. In 2018, more than ten years later, the mission grew to more than 20.000. AMISOM not only shows the commitment of

---

[19]See Article 6 of the Protocol Relating to the Establishment of the Peace and Security Council of the African Union. http://www.peaceau.org/uploads/psc-protocol-en.pdf, accessed August 26th, 2018.

[20]Constitutive Act of the African Union. https://au.int/sites/default/files/pages/32020-file-constitutiveact_en.pdf, accessed August 26th, 2018.

[21]Although most external non-African analysts are a kind of obsessed with peace and security it is only one out of 12 fields of engagement which the AU towards a peaceful, prosperous and integrated Africa. See https://au.int, accessed August 26th, 2018.

the AU and its member states to provide "African solutions to African problems", it shows the shortcomings too.[22]

The view of the AU is however much broader and not only focussed on the narrow perspective on security. The Union has a vision on development and human security which is reflected in its African Agenda 2063 adopted in 2015. It amalgamated almost all African and international frameworks, programmes and declarations into African Agenda 2063, which should contribute to the economic, social, political, scientific and cultural development of the continent. It includes inter alia the New Partnership for Africa's Development (NEPAD), the Lagos Plan of Action, the African Economic Community ("Abuja treaty"), the Minimum Integration Programme, and the Comprehensive Africa Agricultural Development Programme (CAADP). The grassroots level of stakeholders has to have and impact as does 35 strategic and action plan of the RECs and states. The study of global mega trends and situational analysis contribute to Agenda 2063. Although it predates international agreements like the global Agenda 2030 and the Sustainable Development Goals (SDGs) these arrangements complement the continental plan. With the adoption of the First Ten Year Implementation Plan (FTYIP) in 2015 the realization has started.[23] As these programmes are designed to contribute to human security in the long run, in contrast to narrow security perspectives the effects might not be measured and made visible until the end of FTYIP.

6   The Contributions of the UN

The UN has a large influence on security in Africa. It is still the largest contributor to peace and security missions and operations on the continent, financially as well as with troops. In 2018 seven ongoing operations led by the department of peacekeeping operation show the impact and the importance a UN engagement.[24] The impact the UN has is undoubtedly positive aspects, but there are negative too.[25] Violations and abuses of human rights and other forms of misconduct lead the list of atrocities conducted by UN personnel. Focussing on personal misbehavior of individuals may distract the view on the so-called bigger picture.

---

[22]There are plenty of papers analyzing the AU in Somalia, its challenges and possible solutions. Funding is just one of it. With all the critique at hand the improvements the AU made since the establishment of AMISOM should not be forgotten. See e.g. Williams, Paul D., Strategic Communications for Peace Operations: The African Union's Information War Against al-Shabaab, Stability. International Journal of Security and Development. 7(1), p.3, https://www.stabilityjournal.org/articles/10.5334/sta.606/, accessed August 26th, 2018 or Mahmood, Omar S and Ndubuisi, Christian Ani, Impact of EU funding dynamics on AMISOM, East Africa Report 16, December 2017, accessed August 26th, 2018.

[23]About Agenda 2063, https://au.int/agenda2063/about, accessed August 27th, 2018.

[24]See the Department of Peacekeeping operations, https://peacekeeping.un.org/en/where-we-operate, accessed August 26th, 2018.

[25]Akonor, Kwame. UN Peacekeeping on Africa. A critical Examination and Recommendations for Improvements, Springer 2017.

The impact huge longterm missions like in the Democratic Republic of the Congo or in South Sudan are often questioned and questions about their failure are raised. African colleagues like Manyok Chol David argue that "it is undeniably true that African nations are parties to UN conventions … however … the African conflicts should be solved by Africans … and UN should just embark on empowering African institutions … to manage regional conflicts".[26] Critical analyses come from humanitarian organizations as well, which question the concepts like the Responsibility to Protect (R2P) or Protection of Civilians (POC) developed during the last decades: "Despite mandates including explicit calls to 'protect civilians', many of these missions were seen as lethargic and ineffective."[27] Haysom and Pederson argue that Chap. 7 missions like UNAMID and UNMISS "are more known for their inability to stand up government forces than for their robustness."[28] Nevertheless, as long as African organizations are logistically and financially dependent on external actors it is highly unlikely that the UN will just mandate African engagement without playing a significant role on the ground.

The UN has however a large impact on the development of the broader view on security. The concept human security of human security as developed by the UN will have an impact on all African countries, although as for the RECs a reliable and valid measurement will only be possible in years or even decades. Having in mind the UN definition of human security[29] it is more a framework or guiding principles than a short-term intervention with a specific purpose.

7 Conclusions

The mechanisms of tackling security issues have to be improved as rapid response has to. Security structures like the RECs and the AU can therefore not be static, but have to adopt different approaches over time to become more dynamic in order to keep the pace with the challenges. Especially since borders between states and RECs are not taken into consideration by non state actors. On a short-term basis a narrow view on security should help to resolve and avoid violent conflicts while in the long run only a broad view dealing with all aspects of human security may lead to a lasting and stable political environment.

---

[26]See e.g. Manyok Chol David, Why UN Missions fail in Africa, Bor Globe Network, August 15th, 2018, http://www.borglobe.com/why-un-missions-fail-in-africa/, accessed August 27th, 2018.
[27]Haysom, Simone and Pedersen, Jan, Robust peacekeeping in Africa: the challenge for humanitarians, October 2015, https://odihpn.org/magazine/robust-peacekeeping-in-africa-the-challenge-for-humanitarians/, accessed August 27th, 2018.
[28]ibd.
[29]"Human security is an approach to assist Member States in identifying and addressing widespread and cross-cutting challenges to the survival, livelihood and dignity of their people" (General Assembly Resolution 66/290).

Robust peacekeeping has a come under fire for its ineffectiveness, but the alternatives given, if there are alternatives given at all, are not convincing. In general, the measurement of success or failure is tricky. However, it seems to be easier to analyze the impact of narrow perspectives as they should have more immediate results. Human security as a broad perspective on security needs different instruments to show the impact as it covers decades of development and, probably, a less impatient attitude.[30] So are narrow and broad perspectives comparable and should they be compared? The instruments and intentions summarized under the two perspectives serve different tasks and want to achieve different goals. Having this in mind a comparison would not lead to any added value in analysis and in practice.

Another important conclusion has to be drawn: African countries are by no means passive entities without an individual agenda. They, their political and economic elites, have interests, which they try to realize and push through. They are very well aware on how the different systems (UN, AU, RECs) work and try to employ them for the pursuit of their interests. An interesting example provides the UN Force Intervention Brigade in the DRC and a similar intention for South Sudan, which failed. In case of the DRC a major troop contributor had economic interests, while in South Sudan no single country had enough interests to pledge troops. While in DRC it was in the political and economic interest of South Africa[31] to intervene, the idea for South Sudan was watered down and finally rejected.[32,33]

Researchers, diplomats, politicians as well as military planners have to reconsider the way how responsibilities are divided. Instead of a regional/geographical approach most probably the future will see a more problem or challenge centered approach. However, this will not only be applicable to African challenges, but has to be a global approach.

Author Biography

Gerald Hainzl has been a Senior Researcher at the Institute for Peace Support and Conflict Management at the National Defense Academy in Vienna since 2004. His research focusses on conflicts and international crises management in Africa. His special interests are non-state actors and their construction of identity as well as making conflict research available to a wider public. He authored many papers and articles and developed a Political Advisor course for missions and operations in Africa together with the Kofi Annan International Peacekeeping Training Center in Accra, Ghana.

---

[30]Rosling, Hans, Rosling, Ola and Rosling Rönnlund, Anna, Factfulness: Ten Reasons We're Wrong About The World—And Why Things Are Better Than You Think, Hodder & Stoughton Ltd, UK, 2018.

[31]South Africa to take up FIB restructuring with UN Department of Peacekeeping, http://www.defenceweb.co.za/, accessed August 27th, 2018.

[32]Nicols, Michelle, U.N. ready to work on proposal for combat force in South Sudan, July 13th, 2016, Reuters World News, https://www.reuters.com/article/us-southsudan-security-un-idUSKCN0ZT1QA, accessed August 27th, 2018.

[33]Tanza, John, UN Official: Peacekeepers in South Sudan Not Intervention Force, VOA, December 14th, 2017, https://www.voanews.com/a/south-sudan-unmiss-mandate/4163740.html, accessed August 27th, 2018.

Space Applications Supporting Justice

Annette Froehlich (iD)

Abstract

Goal 16 of the Sustainable Development Goals promotes "just, peaceful and inclusive societies". This presupposes fair and equal judgements of justice. However, judgements can only be objective if they are based on correct information. Satellite data may assist in this regard, especially in regard to territorial conflicts, as is the case for wide areas such as on the Africa continent. In addition, the International Court of Justice has also accepted satellite data in other cases, for example to prove the installation of weapons, to monitor the equitable implementation of its judgments or to prove the location of settlements of local populations. More recently, satellite data have also become known in other fields, such as the detection or proof of the violation of human rights.

1 Goal 16 SDG and the Promotion of Just, Peaceful and Inclusive Society

The Goal 16 of the Sustainable Development Goals (SDG) is entitled: "Promote just, peaceful and inclusive societies" and is "dedicated to the promotion of peaceful and inclusive societies for sustainable development, the provision of access to justice for all, and building effective, accountable institutions at all

A. Froehlich (✉)
University of Cape Town, Rondebosch, South Africa
e-mail: Annette.Froehlich@uct.ac.za

© Springer Nature Switzerland AG 2019   265
A. Froehlich (ed.), *Embedding Space in African Society*, Southern Space Studies,
https://doi.org/10.1007/978-3-030-06040-4_16

levels".[1] The following "facts and figures" paragraph elaborates further that "among the institutions most affected by corruption are the judiciary and police. The rule of law and development have a significant interrelation and are mutually reinforcing, making it essential for sustainable development at the national and international level."[2] Therefore, it is enumerated among the targets of this Goal 16 that it should "promote the rule of law at the national and international levels and ensure equal access to justice for all" and to "develop effective, accountable and transparent institutions at all levels".[3]

However, this Goal 16 was seen as controversial during its drafting period. "The mention of justice, governance and peaceful societies in the SDGs is seen as an important step, but one that will pose many challenges. Secretary General Ban Ki-Moon has put his support behind the inclusion of justice as a central pillar for achieving sustainable development. (...) The UN member states are set to adopt the final SDGs agenda at the Special Summit on Sustainable Development in New York in September, but the inclusion of Goal 16 in the final SDGs is not certain. Surprisingly broad support from a large majority of UN member countries—including a large number of developing countries—suggests that dropping this goal would be difficult at this late stage. While some influential countries are not supportive of Goal 16, others are pushing to reopen debate to set a more ambitious scope for the goal which could further jeopardize its inclusion in the final SDGs agenda."[4]

Moreover, it has to be underlined that justice can only provide for more security and stability if it is based on equitable judgements. Indeed, each community requires for its good and peaceful functioning a recognized justice whose judgements are considered as equitable to ensure sustainable development. Thus, this leads to lasting solutions by reducing tensions among habitants and populations.[5] This is of particular importance as on the Sustainable Development Knowledge Platform[6] it is indicated as "progress of goal 16 in 2017" among other aspects: "Violent conflicts have increased in recent years. Progress promoting peace and justice, together with effective, accountable and inclusive institutions, remains uneven across and within regions."[7]

---

[1]Goal 16: Promote just, peaceful and inclusive societies, https://www.un.org/sustainabledevelopment/peace-justice/ (accessed 10 July 2018).
[2]See Footnote 1.
[3]See Footnote 1.
[4]The World Bank, Governance for Development, Justice proposed for sustainable development goals, Submitted by Heike Gramckow, co-authors: Nicholas Menzies, 02/09/2015, http://blogs.worldbank.org/governance/justice-proposed-sustainable-development-goals
[5]Peace, Justice, And Strong Institutions: Why They Matter, p.2, https://www.un.org/sustainabledevelopment/wp-content/uploads/2017/01/16-00055p_Why_it_Matters_Goal16_Peace_new_text_Oct26.pdf (accessed 10 July 2018).
[6]Sustainable Development Knowledge Platform, https://sustainabledevelopment.un.org/sdg16 (accessed 10 July 2018).
[7]See Footnote 6.

Therefore, if a judgement is not considered as equitable by its local populations or concerned persons, it will not provide for stability among the community leading to conflicts and more contentious cases. Judgements will not be accepted or no effort will be made to get them implemented. "Lack of access to justice means that conflicts remain unresolved and people cannot obtain protection and redress. Institutions that do not function according to legitimate laws are prone to arbitrariness and abuse of power, and less capable of delivering public services to everyone. To exclude and to discriminate not only violates human rights, but also causes resentment and animosity, and could give rise to violence." [8]

Nevertheless the question remains, how can a judgement be considered as fair, especially if the circumstances presented to the judges are not reflecting the true situation? Thus, satellite data may support governments and justice in their work for a more secure environment. This may be the case in large area like on the African continent as satellite imageries can support in monitoring the environment and protect against illegal actions (such as illegal fishing/hunting, environmental pollution etc.). Therefore, the International Court of Justice (ICJ) admitted a long time ago (since the 1990s) the use of satellite data for territorial delimitation questions (result of wide geographical surface with low institutionalized infrastructure). These ICJ cases related to African countries will be analyzed in the first part. In the second part, ICJ cases related to litigious issues outside of Africa will be presented to highlight further possibilities to use satellite data in trials. In conclusion, new fields in which satellite data can be used to prove human rights violations will be analysed.

2 Satellites Imageries in ICJ Cases Related to Territorial Delimitation Questions on African Continent

By analyzing the judgements of the ICJ, it can be revealed how important satellite data are in legal proceedings and how they can support judges in finding a fair judgement. Concerning territorial delimitation disputes, satellite data were used either to prove the historical evolution of geographical features which are essential to determine the course of the frontier, or to ensure and monitor the implementation of its judgments.

2.1 Satellite Data to Prove the Evolution of a Border River

One of the earliest cases which relied on satellite images is the judgement of the ICJ *on the status of the Kasikili/Sedudu Island*[9] between Botswana and Namibia[10]

---

[8]See Footnote 5.
[9]Kasikili is the Namibian name, Sedudu the Botswana name for the island.
[10]Case concerning Kasikili/Sedudu Island (Botswana/Namibia), Judgement of 13 December 1999, I.C.J. Reports 1999.

were the Court admitted the use of satellite images as proofs. Those satellite data were very useful to reveal the historical evolution of the "main channel" of a river. This was indispensable to resolve a territorial delimitation question related to this Island of 3,5 sq km, located in the Chobe River and being frequently subject to flooding of several months' duration. The litigious question submitted to the ICJ concerned the right interpretation of the Treaty between Great Britain and Germany (The Anglo-German Agreement of 1 July 1890) in which the former colonial powers Germany (for the contemporary country of Namibia, formerly German Southwest Africa) and the United Kingdom (for the contemporary country of Botswana, previously known as Bechuanaland) settled their geographic interests. The dispute arose due to the imprecise wording of the agreement concerning the northern boundary between Namibia and Botswana.

Therefore the ICJ focused on the definition of the term "centre of the main channel" as Art. III of this Anglo-German Treaty made reference to "centre of the main channel" (the German version of that treaty to "Thalweg des Hauptlaufes") of the river Chobe. However, the question emerged as to how to determine the main channel as the treaty itself did not precisely identify it. Since a Joint Team of Technical Experts did not reach a conclusion, the two countries decided to over-come their difference by peaceful means in accordance with the principles of both the Charter of the United Nations and the Charter of the Organization of African Unity. As a consequence, at the Summit Meeting held in Harare, Zimbabwe, on 15 February 1995, President Masire of Botswana and President Nujoma of Namibia agreed to submit the dispute to the International Court of Justice for a final and binding determination. Thus, the two countries sent a joint letter to the ICJ to define the boundary between Namibia and Botswana around Kasikili/Sedudu Island and its legal status on the basis of the Anglo-German Treaty of 1st July 1890 and the rules and principles of international law. In regard to the "main channel" of the river Chobe, Botswana claimed that it was the channel running north of Kasikili/Sedudu Island, while Namibia insisted that the channel ran south of the island.

In this context, it has to be mentioned that this Anglo-German Agreement was not a classic boundary treaty. Therefore, the ICJ declared: "While the treaty in question is not a boundary treaty proper, but a treaty delimiting spheres of influence, the Parties nonetheless accept it as the treaty determining the boundary between their territories. The major concern of each contracting party was to protect its sphere of influence against any intervention by the other party and to obviate any risk of future dis-putes."[11] Moreover, the ICJ concluded: "the contracting powers, by opting for the words 'centre of the main channel', intended to establish a boundary separating their spheres of influence even in the case of a river having more than one channel. They possessed only rudimentary information about the Chobe's channels. If they knew that such channels existed, their number, features, navigability, etc., and their relative importance remained unknown to them. This situation explains the method adopted to define the southern boundary of the Caprivi Strip."[12]

---

[11]Pt. 33.
[12]Pt. 41.

Indeed, the ICJ stated already in its Temple of Preah Vihear case: "There are boundary treaties which do no more than refer to a watershed line, or to a crest line, and which make no provision for any delimitation in addition"[13]. Moreover, the Court added in the same judgement that this was "an obvious and convenient way of describing a frontier line objectively, though in general terms".[14] Consequently, the ICJ concluded that "in the present case, the contracting parties employed a similar approach."[15]

Concerning the litigious aspect of this treaty itself, Namibia underlined in its statement that it has a prescriptive title to Kasikili/Sedudu island as the Masubia population of the Caprivi Strip used this island for agriculture purposes and founded with its community a well-organized village, were sovereign jurisdiction was exercised. These facts were fully known by the Bechuanaland authorities in Kasane, just across the river in Botswana. Therefore, Namibia concluded that "the continued control and use of Kasikili Island by the people of the Eastern Caprivi, the exercise of jurisdiction over the Island by the governing authorities in the Caprivi Strip, and the continued silence of those on the other side of the Chobe … confirm the interpretation of the Treaty … [whereby] Article III … attributes Kasikili Island to Namibia".[16] However, Botswana did not dispute that people from the Caprivi Strip used the island for agriculture purposes, but pretended that their use was of "sporadic nature",[17] in the same way as this was done by people from the Bechuanaland. Botswana even contested that there was a permanent settlement or a village on this island.

In regard to these litigious facts, the Court expressed its view that the settlement of the Masubia tribespeople of the Kasikili/Sedudu Island should constitute a "subsequent practice in the application of the [1890] treaty which establishes the agreement of the parties regarding its interpretation".[18] The Court follows that in order "to establish such practice, at least two criteria would have to be satisfied: first, that the occupation of the Island by the Masubia was linked to a belief on the part of the Caprivi authorities that the boundary laid down by the 1890 Treaty followed the southern channel of the Chobe; and second, that the Bechuanaland authorities were fully aware of and accepted this as a confirmation of the Treaty boundary".[19] Therefore, the Court rejected the argument of Namibia as the use of the Island by the Masubia tribespeople "had, in recent years, become infrequent"[20] (according to the joint survey report on the Chobe River drawn up by South African and Botswanan experts on 15 July 1985 in the context of discussions on the location of the boundary around Kasikili/Sedudu Island). Therefore, the Court

---

[13]Case concerning the Temple of Preah Vihear case (Cambodia v. Thailand), Merits, Judgement of 15 June 1962, p. 32.
[14]See Footnote 11.
[15]Case concerning Kasikili/Sedudu Island, pt. 43.
[16]Pt. 71.
[17]Pt. 72.
[18]Pt. 73.
[19]Pt. 74.
[20]See Footnote 19.

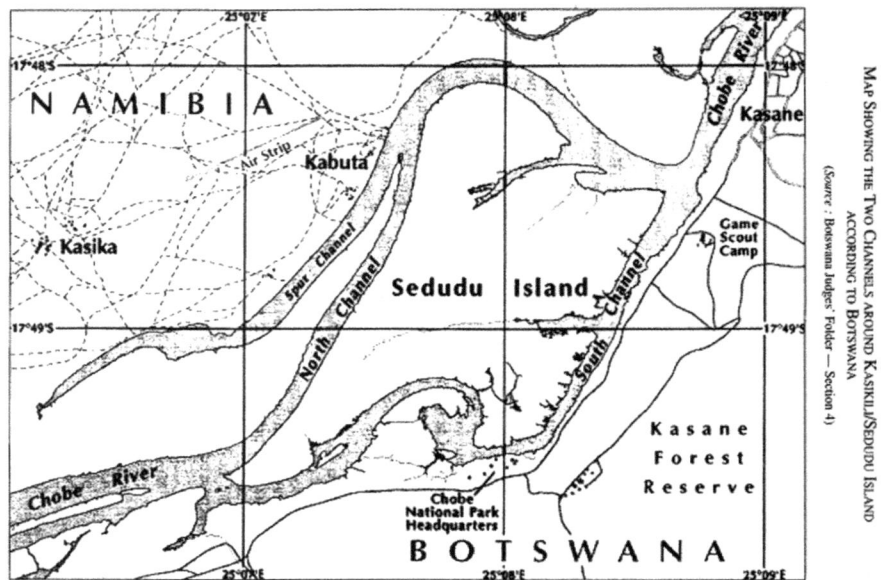

Fig. 1 Map Showing the two channels around Kasikili/Sedudu Island according to Botswana. (Pt. 35)

concluded "that the peaceful and public use of Kasikili/Sedudu Island, over a period of many years, by Masubia tribesmen from the Eastern Caprivi does not constitute 'subsequent practice in the application of the [1890] treaty'."[21]

Consequently, the ICJ had to focus on the geographical features to determine the legal statute of the disputed island. Due to aforementioned lack of definition of the term "main channel" in the Anglo-German Treaty, the Court itself proceeded to determine the main channel of the Chobe River around the Island and took into consideration the depth and the width of the channel, the flow (i.e., the volume of water carried), the bed profile configuration and the navigability of the channel. After considering the figures submitted by both parties, as well as surveys carried out at different periods, the Court concluded that the boundary between Botswana and Namibia around Kasikili/Sedudu Island followed the line of deepest soundings in the northern channel of the Chobe and that the Island formed part of the territory of Botswana (Fig. 1).

It has to be highlighted that Botswana has founded its arguments on scientific evidence[22] "that the northern channel, by reason of the greater depth, width and bed profile, is the navigable channel capable of carrying the greater flow, and hence is

---

[21]Pt. 75.

[22]Case concerning Kasikili/Sedudu Island (Botswana/Namibia), Reply of Botswana, Vol. 1, Nov. 1998, Chap. 6 "The Scientific Evidence", pt. B, Botswana's Case based on the Scientific Evidence, pt. 298.

the main channel of the Chobe River in the vicinity of Kasikili/Sedudu Island. Botswana's case is supported by all the official surveys carried out on the spot, and by scientific evidence based on geomorphology, hydrology and fieldwork, as well as the 70-year series of aerial photographs taken between 1925 and 1997, and the satellite imagery of 1975, 1995 and 1996."[23]

Due to the evidence material, this case gained of particular importance as "much of this material is either new or was previously inaccessible. Scientists play an increasingly important role in the resolution of disputes, even at international level"[24] and among these scientific sources satellite data played a crucial role.

2.2 Satellite Data for Frontier Disputes

However, even knowing the main channel of a river may lead to conflicts as it may occur that it has shifted over the past, a situation which may raise further frontier disputes. This also happened in the case between *Burkina Faso and Republic of Mali concerning their frontier dispute*[25], a border conflict due to the fact that the riverbed had geographically shifted from when the border was set during colonial period. During the proceeding, various maps and data, among them satellite imageries, were presented by the parties to the ICJ to settle the demarcation dispute. The parties even agreed to consult three experts which should assist in the demarcation process.

In the course of this process, the Court underlined that in order to secure respect for the territorial boundaries, the principle of the intangibility of frontiers inherited from colonization time and the principle of *uti possidetis juris*, meaning that a legal title accords pre-eminence over effective possession, applies in this case. In addition, the ICJ clarifies that boundaries during colonization may have been only delimitations between different administrative entities which belonged all to the same sovereign, but which however may have become international frontiers, as this was the case in the dispute between Mali and Burkina Faso.

In this context satellite imageries were of help to situate or to re-locate the riverbed as it has changed its course over decades since the decolonization period. Indeed, satellite data could show their merits as on both sides the local populations may feel justified to argue that their settlement were on one or another side of the river. From the daily perspective of living on the ground, they could not have recognized a change of the riverbed which may have shifted slowly over decades (especially in wide geographically areas with low or rural infrastructure).

Therefore, satellite data can provide an overview of the whole area and, in combination with data from the past, a more reliable impression of the concerned topographic situation allowing a better evaluation of the circumstances in border

---

[23]See Footnote 22.
[24]W.J.R. Alexander, Science, history and the Kasikili Island dispute, http://webworld.unesco.org/water/wwap/pccp/cd/pdf/educational_tools/course_modules/reference_documents/sharinginternwater cases/sciencehistory.pdf.
[25]Frontier Dispute between Burkina Faso and Republic of Mali, I.C.J. Reports 1986, p. 554.

delimitations disputes. However, the Court questioned the technical reliability of the maps during the process, the neutrality of its sources and the multiple ways of interpretation of the data.[26]

2.3 Satellite Data to Solve Human Settlements

A further case related to territorial delimitation also emanating from the African continent is known as *"Land and Maritime Boundary between Cameroon and Nigeria"*.[27] Cameroon brought this border dispute between the two countries to the ICJ with respect to the question of sovereignty over Bakassi Peninsula containing vast oil resources. Thus, the Court was requested to determine the course of the maritime frontier between the two States. However, this case is of special interest as satellite imageries were not only presented during the proceedings, but were especially of utmost importance after the ICJ delivered its judgment to monitor its implementation.

Indeed, the ICJ decided—based on the Anglo-German Agreement of 11 March 1913—that the sovereignty over the Bakassi Peninsula belongs to Cameroon as the boundary is constituted by the *Thalweg* of the River Akpakorum (Akwayafe). However, in regard to the delimitation of the maritime boundary between Cameroon and Nigeria, a delimitation was not so easy to establish as these two countries share a common border of about 1700 km (the boundary extends from Lake Chad in the north to the Bakassi Peninsula in the south) with coastlines adjacent and washed by the waters of the Gulf of Guinea. As the ICJ judgment seemed difficult to implement, a Cameroon-Nigeria Mixed Commission was established with the assistance of the United Nations to ensure a peaceful implementation of this judgment.[28] However, the straddling settlements along the boundary posed problems in view of a just implementation of the Court's judgment.[29] This was due to the fact that water variations over the time lead to changes of settlements as the local population just followed the receding waters and cultivate the arable land it leaved behind.

In consequence, satellite data revealed that both countries had villages across the newly delimited boundary lines as the Court's judgment has not considered the lives and circumstances of people. Therefore, the Mixed Commission was asked to find a solution in order to avoid further conflicts. Satellite images were acquired

---

[26]Pt. 56.
[27]Land and Maritime Boundary between Cameroon and Nigeria (Cameroon v. Nigeria: Equatorial Guinea Intervening), I.C.J. Report 2002
[28]Annette Froehlich, Space Related Data: From Justice To Development, IAC 2011 (IAC-11-E7.3.1), p. 5.
[29]Cf. *Gbenga Oduntan*, The Demarcation of Straddling Villages in Accordance with the International Court of Justice Jurisprudence: The Cameroon–Nigeria Experience, Chinese Journal of International Law, 2006, Vol. 5, Issue 1, pp. 79–114, http://chinesejil.oxfordjournals.org/content/5/1/79.full (accessed: 15.6.2018)

by UN cartographic experts to draw up the final demarcation maps[30] to settle this dispute in order to guarantee a peaceful environment allowing stability and security in this area.

3 Further ICJ Cases with Reference to Satellite Data for a Better Overview

Furthermore, the ICJ accepted satellite data during various proceedings which were not related to the African continent, but are of special interest as they highlight the importance and utility of satellite images for trials.[31] Indeed, they were not only useful for territorial delimitation questions to resolve geographical questions, but also to demonstrate environmental impacts of certain actions or to prove installations of military equipment.

3.1 Satellite Data for Maritime Delimitation

The following several cases are analysed in order to show how useful satellite data may be to resolve conflicts related to maritime delimitation issues. In this regard, the case *"Territorial and Maritime Dispute between Nicaragua and Honduras in the Caribbean Sea"*[32](2007) should be firstly mentioned. The Court was asked to adjudge on sovereignty over four islands located in the disputed area. In this procedure, Honduras showed a "set of satellite photos of the river mouth in seven different years from 1979 to 2011"[33] proving the migration of the islands. Based on these satellite images, it became obvious that the islands have formed in the mouth of the Rio Coco, as the river drops its sediment.

---

[30]"The Nigerian Boundary Commission reported that, as of January 2006, implementation of the ICJ judgment was progressing. 'Both countries [have] secured the technical assistance of the UN to undertake the field work ... [and] have secured the latest satellite imagery of the border area 30 km in Nigeria and 30 km in Cameroon.' With satellite mapping, a technical team of Nigerian, Cameroonian, and UN officials reportedly commenced intense cartographic demarcation work in the field in accordance with the judgment." *Aloysius P. Llamzon,* Jurisdiction and Compliance in Recent Decisions of the International Court of Justice, EJIL 2007, p. 838, http://www.ejil.org/pdfs/18/5/250.pdf (accessed: 15.6.2018).

[31]See further: Annette Froehlich, The Impact of Satellite Data used by High International Courts like the ICJ (International Court of Justice) and ITLOS (International Tribunal for the Law of the Sea), Proceedings of the International Institute of Space Law 2012, Den Haag 2013, pp. 471–483; Annette Froehlich, Space Related Data: From Justice To Development, Proceedings of the International Institute of Space Law 2011, Den Haag 2012, pp. 221–227.

[32]Territorial and Maritime Dispute between Nicaragua and Honduras in the Caribbean Sea (Nicaragua v. Honduras), I.C.J. Reports 2007.

[33]ICJ, Public sitting 15 March 2007 in the case concerning Maritime delimitation between Nicaragua and Honduras in the Caribbean Sea (Nicaragua v. Honduras), Verbatim record, p. 48, pt. 4 (a).

Also in the case *"Maritime Delimitation and Territorial Questions between Qatar and Bahrain"* (2001),[34] both parties used satellite imageries to underline their well-founded allegations. The legal regulations were clear, but it was difficult to apply them to the concrete situation in the relevant area as necessary position points could not clearly be defined. Indeed, the debate related to the measuring of sea-level rise due to the difference in the choice of mean sea level datum and variations in tidal heights. Therefore, satellite data were of great support to achieve a higher level of accuracy in measuring the sea-level in order to apply legal norms of delimitation fixed in former agreement.

Furthermore, the case *"Maritime delimitation in the Black Sea"* (2009)[35] related to questions around the continental shelf and the exclusive economic zones between Romania and Ukraine, a region which was of particular interest due to its oil and natural gas deposits under its seabed. Hereby the geographic position of the Serpent Island was of utmost importance as its status dramatically affects the maritime frontier line between the two countries. In this context Romania's argumentation was based (among others) on publicly available satellite images depicting that "Serpents' Island is a small, uninhabitable feature which is in no sense integrated with the mainland coast, from which it lies at more than 20 nm. Eloquent in this respect is the satellite picture of the area which shows Serpents' Island lying out at sea, by no means integrated with the mainland".[36]

In addition, during the procedure on the *"Sovereignty over Pedra Branca/Pulau Batu Puteh, Middle Rocks and South Ledge"*,[37] a territorial dispute between Malaysia and Singapore in regard to various islets at the eastern entrance to the Singapore Strait, Singapore submitted satellites images proving that the Island of Pedra Branca does not form part of the "Romania Island group" geographically, but is an independent feature.

Also in the case between Indonesia/Malaysia concerning the Status of the two very small islands *"Pulau Ligitan and Pulau Sipadan"*,[38] Malaysia stated in its memorial "[T]the location of Ligitan and Sipadan, and their relation to the other places mentioned here, can be seen from the satellite images of the region".[39]

---

[34]Maritime Delimitation and Territorial Questions between Qatar and Bahrain (Qatar v. Bahrain), I.C.J. Report 2001.

[35]Maritime Delimitation in the Black Sea (Romania v. Ukraine), I.C.J. Reports 2009.

[36]Reply by Romania Pt. 8.18.

[37]Sovereignty over Pedra Branca/Pulau Batu Puteh, Middle Rocks and South Ledge (Malaysia/Singapore), I.C.J. Reports 2008.

[38]Sovereignty over Pulau Ligitan and Pulau Sipadan (Indonesia/Malaysia), I.C.J. Reports 2002.

[39]Case concerning sovereignty over Pulau Ligitan and Pulau Sipadan, Memorial of Malaysia, 2.11.1999, p. 19, pt. 3.22, http://www.icj-cij.org/docket/files/102/8560.pdf, furthermore reply of Malaysia, 2.3.2001, p. 13, pt. 2.14, fn. 34, http://www.icj-cij.org/docket/files/102/8566.pdf.

3.2 Satellite Data to Demonstrate Environment Impacts

Moreover, the increasing importance of satellite data in the context of environmental impacts has to be underlined. This was set out in the case *"Certain Activities Carried out by Nicaragua in the Border Area"*[40] (2015). Indeed, Costa Rica introduced proceedings against Nicaragua due to alleged two separated incidents, the occupation and use of its territory by Nicaragua's army and the damaging of Costa Rica's environment by constructing a canal across Costa Rican territory and hereby seeking to divert the flow of the San Juan River.[41] For the dredging, trees were felled and sediments from the dredging work were deposited on Costa Rican territory. The dredging of the San Juan River[42] "seriously affected the flow of water to the Colorado river of Costa Rica, and will cause further damage to Costa Rican territory, including the wetlands and national wildlife protected areas located in the regions".[43] Additionally, Costa Rica accused Nicaragua of causing serious impacts on the environment, habitat and especially on the wetlands as a result of these activities. Therefore, Costa Rica referred "to a report of 4 January 2011 drawn up by the Operational Satellite Applications Programme of the United Nations Institute for Training and Research ("UNITAR/UNOSAT report") relating to the geomorphological and environmental changes likely to be caused by Nicaragua's activities in the border region".[44] Nicaragua however stated that the works on the canal were necessary due to the fact that it became obstructed over the years by the sedimentation of its river bed. The works caused no harm to the environment and were necessary to make the canal more navigable for small vessels, an obligation for the sovereign of the river under international law.[45,46] Even though it is a territorial conflict, the subject of the clean-up operation and its environmental impact has been especially evidenced by various maps and satellite photographs. In consequence, the Court considered it as proven that Nicaragua has violated Costa Rica's territorial sovereignty and thus has to compensate Coast Rica for its material damage.

3.3 Satellite Data to Prove Military Weapon Installations

The case *"Oil Platforms"* which opposed the Islamic Republic of Iran to the United States of America (2003)[47] demonstrates that satellites data may be of utmost importance not only to reveal the use of weapons, but also that the geographical

---

[40]Certain Activities Carried Out by Nicaragua in the Border Area (Costa Rica v. Nicaragua) *and* Construction of a Road in Costa Rica along the San Juan River (Nicaragua v. Costa Rica), Judgment, I.C.J. Reports 2015, *p. 665.*

[41]Certain Activities Carried Out by Nicaragua in the Boarder Area (Costa Rica v. Nicaragua), Request for the indication of provisional measures, 8 March 2011, Nr. 6.

[42]Nr. 3.

[43]Nr. 5.

[44]Nr. 33.

[45]Nr. 38.

[46]Nr. 40.

[47]"Oil Platforms" (Islamic Republic of Iran v. United States of America), I.C.J. Reports 2003.

features and the consistence of the ground was allowing the installation of a solid weapon system for cruise missiles. In the context of this process, the Islamic Republic of Iran instituted the proceeding by claiming that the destruction of three offshore oil platforms caused by several warships of the United States Navy (in the course of various incidents during 1987 and 1988) constituted a fundamental breach of international provisions. In its counter-memorial, the United States of America alleged however that it was Iran who violated international regulations by attacking vessels in the Gulf. Therefore, the oil platforms were destroyed as Iran used them to carry out attacks against US and other neutral ships during the Iran-Iraq war. Moreover, the US submitted satellite data by stating that this "photographic evidence and expert testimony will squarely refute Iran's claim that it did not maintain missile sites in the Faw area, including its claim that the Faw was composed 'almost entirely' of marshland, and was therefore incapable of sustaining missile sites".[48] Furthermore it is stated: "The evidence shows how Iran carried out deadly armed attacks on U.S. vessels. Eyewitness accounts of Iran's missile attack on the U.S.-flag tanker *Sea Isle City* on 16 October 1987, analysis of missile fragments, and satellite imagery help to demonstrate Iran's responsibility for that attack."[49] It should be underlined that the US national security considerations preclude the US from submitting the original data, therefore their resolution was reduced to allow the public dissemination of those satellite imageries.

4 New Fields: Use of Satellite Data to Prove Human Rights Violations

In more recent times, satellite imageries were even presented to the ICJ to prove the violation of human rights, for example in the case Georgia versus. Russian Federation,[50] but also in relation to the African continent, various incidents led to the use of satellite imageries to prove human rights violations or to use them as tool of investigation.

In this context it has to be highlighted that satellites images published in an Amnesty report were useful to the Zimbabwean Lawyers for Human rights in their presentation to the African Court on Human and People's Rights to demonstrate the housing demolitions in Zimbabwe.[51] Furthermore, satellite imagery can be

---

[48]Counter-memorial and counter claim submitted by the United States of America, 23 June 1997, pt. 1.75.
[49]Counter-memorial and counter claim submitted by the United States of America, 23 June 1997, pt. 1.05.
[50]Application of the International Convention on the Elimination of All Forms of Racial Discrimination (Georgia v. Russian Federation), Preliminary Objections, Judgment, I.C.J. Reports 2011, p. 70.
[51]Ana Cristina Núñez M., Admissibility of remote sensing evidence before international and regional tribunals, Innovations in Human Rights Monitoring, Working Paper, August 2012, p. 16, https://www.amnestyusa.org/pdfs/RemoteSensingAsEvidencePaper.pdf (accessed 15 July 2018).

extremely useful to monitor wide rural areas. Despite the Ethiopian government's announcement to relocate rural people to improve their lives, Human Rights Watch questioned it by presenting satellites imagery series taken between 2009 and 2011 revealing that "villagers living within the Gambella region of Western Ethiopia have been relocated to smaller and less desirable plots of land, possibly to make way for large foreign-owned commercial farms".[52]

Satellite date may also provide evidence of war crimes or serve as an efficient investigation instrument. As an example, genocide in Darfur[53] and the unlawful attack against the peacekeepers of the African Union can be mentioned. Based on satellite data, the damage on the MGS Haskanita military base in northern Darfur in September 2007 could therefore be demonstrated along pre-attack and post-attack satellite imageries.[54] But also various proceedings have been initiated by the International Criminal Court (ICC) in the Democratic Republic of the Congo (DRC) on crimes committed in the DRC during the war and conflicts in its aftermath.[55]

Finally, it has to be highlighted that the Permanent Court of Arbitration (PCA) in The Hague also admitted satellite imagery as evidence, even if they are emanating from commercial satellites. The Ikonos high-resolution satellite imagery supported the claims of Eritrea submitted in 2002 in a report entitled "Ethiopia's Violations of International Law Arising From Its Attacks on and Occupation of the Central Zone of Eritrea."[56] The admission of commercial satellite data is a further valuable step in the use of satellite imagery in trials.

5 Conclusion: New Fields for Experts and Regulations

In conclusion, it can be stated that satellite data were already highly useful in ICJ trials due to the fact that satellite imageries can provide factual data. Especially in combination with satellite data banks of the past, they enable a better evaluation of

---

[52]Amy Maxmen, AAAS Geospatial Report: Ethiopian "Villagization" Policy is Displacing Farmers in Gambella Region, 17 January 2012,
https://www.aaas.org/news/aaas-geospatial-report-ethiopian-villagization-policy-displacing-farmers-gambella-region (accessed 7 October 2018).
[53]Russell Schimmer, Tracking the Genocide in Darfur: Population Displacement as Recorded by Remote Sensing, Genocide Studies Working Paper No. 36, https://gsp.yale.edu/sites/default/files/files/GS36.pdf (accessed 7 October 2018).
[54]ICC, Situation in Darfur, the Sudan in the Case of the Prosecutor v. Bahar Idriss Abu Garda, 24 September 2009, ICC-02/05-02/09.
[55]See Democratic Republic of the Congo, Case: *The Prosecutor v. Thomas Lubanga Dyilo; The Prosecutor v Germain Katanga and Mathieu Ngudjolo Chui, The Prosecutor v. Bahar Idriss Abu Garda*, available at http://www.icc-cpi.int.
[56]AAAS Scientific Responsibility, Human Rights & Law Program, Ethiopian Occupation of the Border Region of Eritrea Case Study Summary, https://www.aaas.org/resources/ethiopian-occupation-border-region-eritrea-case-study-summary (accessed 7 October 2018).

the circumstances. As satellite data can depict a more reliable overview of the whole situation, they are leading to fairer and more equal justice for which the Sustainable Development Goal 16 is reaching. As highlighted at the beginning of this analysis, more justice means more stability within every community.

However, Courts may need (more) experts in satellite data in the future in order to appreciate and get a better insight into the reliability of satellite data and its interpretation. This is of particular interest as there are still no uniformed regulations in view of the analysis and interpretation of satellite data. Therefore, the ICJ raises this in its above-mentioned case between Burkina Faso and Republic of Mali concerning the Frontier Dispute,[57] in which the Court questioned the technical reliability of the maps and the various not unified ways of interpretation of the data. Also, in the mentioned *"Oil Platforms"* case, the US took the diligence to present during the oral proceedings a satellite imagery expert to the ICJ explaining and providing more details on the analysis of the submitted satellite data. In order to guarantee the independence and thereby the reliability, it would be recommendable for the International Court of Justice to have also its own in-house expert and not to have to rely on experts presented by the involved parties. This may foster the reliability of the judgements even more. Therefore the International Criminal Court enlarged its in-house expertise. In its manual, based on the outcome of its related workshop, various types of scientific evidence for international investigations are listed such as satellite/remote sensing data, photographic imaginaries/digital and video evidence or forensic techniques. Furthermore, it is stated that "Remote Sensing and Satellite Imaging: Several workshop participants spoke about remote sensing and satellite imagery as potential alternative methods for demonstrating various elements of the crimes. Remote sensing circumvents administrative barriers and security impediments, and minimizes risk to investigators while creating a permanent record of a crime scene or allowing a reconstruction of events over time. Additionally, relative to the cost of sending in investigative teams, remote sensing and satellite imagery are economical ways to establish factual grounds for an accusation in a form that is easily understood in a courtroom setting".[58] Moreover, satellite data enable establishing of the big picture "[B]ecause it can establish patterns over time and across vast distances, scientific evidence can provide judges with a richer understanding of events and their impact on war affected populations. For example, (…) satellite imaging or digital reconstruction of a crime scene can help judges gain a visual and conceptual understanding of the events in question."[59]

---

[57]See Footnote 25.

[58]Beyond Reasonable Doubt, Using Scientific Evidence to Advance Prosecutions at the International Criminal Court, 23–24 October 2012, workshop report, pp. 6–7, https://www.law. berkeley.edu/files/HRC/HRC_Beyond_Reasonable_Doubt_FINAL.pdf (Accessed 15 July 2018)

[59]Beyond Reasonable Doubt, Using Scientific Evidence to Advance Prosecutions at the International Criminal Court, 23–24 October 2012, workshop report, p. 10.

Therefore, a next step should be to elaborate measures or international rules in view of uniform ways of data processing and interpretation as it is the processed data in form of a satellite images which will serve as evidence in courts. By increasingly learning to trust satellite imageries, fewer conflicts will hopefully emerge in consequence, allowing and leading to more development.

Author Biography

Dr. Annette Froehlich is an Honorary Adjunct Senior Lecturer at the University of Cape Town (UCT) (SA) at SpaceLab and scientific expert seconded from the German Aerospace Center (DLR) to the European Space Policy Institute (Vienna). She graduated in European and International Law at the University of Strasbourg (France), followed by business oriented postgraduate studies and her PhD at the University of Vienna (Austria). Moreover, Dr. Annette Froehlich is author of a multitude of specialist publications and serves as lecturer at various universities worldwide in space policy, law and society aspects. Her main areas of scientific interest are European Space Policy, International and Regional Space Law, Emerging Space Countries, Space Security and Space & Culture.

Printed by Printforce, the Netherlands